Praise f
Facing the Climate Emergency

The world is on fire, yet few lift a finger — nobody better understands both the state of the climate crisis and the ways we all delude ourselves into complacency than Margaret Klein Salamon. This book will wake you up, no matter how aware you think you are. Better still: it tells you what to do once you are awake. Read it!

— **David Wallace-Wells**, deputy editor, *New York Magazine*, author, *The Uninhabitable Earth*

We are in very hot water, and we can still do much about it — those two facts set the stage for Margaret Klein Salamon's remarkable account of how you can become a climate warrior.

— **Bill McKibben**, author, *Falter: Has the Human Game Begun to Play Itself Out?*

An epic call to action on, without exaggeration, the greatest threat to mankind in history: the destruction of our livable atmosphere. As I read each page I could slowly feel the ball of fear and stress we live with in the climate crisis come undone and turn into something much more hopeful: purpose.

— **Adam McKay**, Academy Award winning writer and director, *The Big Short* and *Vice*

In the blink of an eye, a movement has been born, and Margaret's voice played an early role in helping to bring it into existence. Now, our movement needs to grow, and quickly. *Facing the Climate Emergency* is an open invitation and a deep spiritual calling to join the millions of people who are waging fierce love to protect humanity from the climate crisis. Drawing on her unique background as a clinical psychologist and a trailblazing climate advocate, Margaret has provided the movement with a complete guide to the inner journey of climate work, from processing our painful emotions to transforming them into greater clarity, connection, and purpose.

— **Varshini Prakash**, co-founder and executive director, Sunrise Movement

A book we can all use at this time of change and instability. With compassion and understanding, *Facing the Climate Emergency* guides us on the inner journey involved in embracing the truths about the climate crisis, processing our upset over its severity, and awakening the love and courage it will take to solve it.

— **Marianne Williamson**, best-selling author, *A Politics of Love*, activist, and spiritual leader

This is the most powerful, honest, and psychologically astute book on climate change I've ever read. If we humans have a collective death wish, we certainly express it in our pervasive climate denial. Here is the antidote.

— **Richard Heinberg**, Senior Fellow, Post Carbon Institute

This book is an inspiring and practical self-help book for the 21st century, challenging readers to overcome denial about the global climate emergency, and honor their grief, fear, and anger, so they can take part in the urgently needed transformation of our society and economy.

— **Molly Brown**, co-author with **Joanna Macy**, *Coming Back to Life: The Updated Guide to the Work that Reconnects*

In Extinction Rebellion we talk about Regenerative Culture. Healing ourselves, and the world. This work is not optional. If we are to play our role in protecting life, each of us must bring our very best qualities to meet this challenge and re-connect to the Earth and to humanity. This wonderful resource skillfully helps us face our most intense emotions about the climate emergency. Read it, use it, and rebel.

— **Gail Bradbrook**, co-founder, Extinction Rebellion

For those feeling overwhelmed by the major crisis of our age, the climate emergency, Margaret Klein Salamon has provided an insightful and very useful guide for individual action.

— **US Congressman, Earl Blumenauer**, sponsor, H. Con. Res. 52, "Declaration of National Climate Emergency"

Margaret Klein Salamon is one of just a handful of people whose work has changed the way the world thinks about climate change. Today everyone is talking "climate emergency", but when she began her work most were in denial about the true scale of the challenge. Such is the power of truth well told. Reading this book will change your life and it could lead you to help change the future – for billions of people's lives.

— **Paul Gilding**, former executive director, Greenpeace International, author, *The Great Disruption*

A must-read, a wake-up call! This brave book takes us into the heart of the climate emergency to rekindle our deep love of life. As guardians of future generations, we must vow to secure our children's future.

— **Raffi Cavoukian, C.M., O.B.C**, singer, author, Raffi Foundation for Child Honouring

Gone are the days when tackling the "climate issue" can be considered the domain of scientists and politicians alone. Addressing the reader directly and personally, Margaret weaves together relatable, moving stories of her journey from clinical psychologist to "climate warrior," with practical insights from psychology greats like Erich Fromm and Abraham Maslow. This authoritative work will serve as an invaluable resource for both mental health professionals and their patients.

— **Nancy McWilliams**, Visiting Full Professor, Rutgers University Graduate School of Applied & Professional Psychology

As fires engulf the planet from the Amazon to Australia and California to the Congo, a global movement calling for climate action and justice has been ignited. In *Facing the Climate Emergency*, Margaret gives us the tools to take action with courage for the immediate change that is needed to protect all life and future generations on our beautiful planet.

— **Leila Salazar-López**, executive director, Amazon Watch

Do you feel anxious about climate change? Powerless to do anything about it? Then you must read this book. Honest, fascinating, deeply inspiring, *Facing the Climate Emergency* will change your life.

— **Genevieve Guenther, PhD.**, founder, End Climate Silence

Margaret rips the cover off denial and inaction on the climate crisis and exposes the truth, unflinchingly, of humanity's plight. No more incremental steps. We must now, collectively, be in "emergency mode" — confronting deeply painful emotions, drawing courage from the power that lies within them.

— **Lise Van Susteren, MD**, general and forensic psychiatrist, board of directors, Physicians for Social Responsibility, Earth Day Network

There's competition for the most important words that can be spoken today, but Salamon's directive "join the climate emergency movement!" has now got to be at the top. Salamon's motivational book, written by someone who has studied the human psyche but has the heart of a compassionate activist, is the leading guide to getting us beyond hopeful complicity and committing to climate action not as a priority but as *the* priority.

— **James Gustave Speth**, author, *America the Possible*, former Dean, Yale School of Forestry and Environmental Studies, co-founder, Natural Resources Defense Council (NRDC)

Facing the Climate Emergency is the ultimate self-help book — because Salamon knows that you and I cannot really help ourselves today unless we do what we can to help heal the earth, including all its beings. And she shows us exactly how to begin that work.

— **David Loy**, Zen teacher, author, *EcoDharma: Buddhist Teachings for the Ecological Crisis*

Margaret Klein Salamon, climate warrior and psychologist extraordinaire, brings her unique experience, wisdom, and compassion to this groundbreaking book, which will help you transform the jagged scraps of your climate pain into the rich compost of climate action.

— **Dr. Peter Kalmus,** NASA climate scientist, climate truth teller, and author, *Being the Change*

As a psychiatrist who has been working on these same issues for the last 7 years, I am deeply grateful and awed by the work Margaret Klein Salamon has done. It details the problems with our thinking, shares the process of getting into emergency mode, and provides practical ways to do that at a level I have not seen elsewhere.

— **Elizabeth Haase, MD**, Climate Psychiatry Alliance, DFAPA

This powerful and eloquent book promises to become the essential guide to climate action. It awakens us to the harsh reality of climate change as a crisis challenging the very foundations of civilization, and offers us practical advice, meticulously laid out, on how to transform arisen fear and grief into a compelling, irresistible drive to action.

— **Ven. Bhikkhu Bodhi**, Buddhist scholar, translator, and activist

Uniquely pairs an honest assessment of the urgency of the crisis with a profound understanding of the personal psychology and social change that it will take to face it. This is one of the most hopeful books I've read in recent years.

— **Jamila Raqib**, executive director, Albert Einstein Institution, and strategic nonviolent action specialist

Salamon's profound yet fast-paced book will help individuals — no matter their age or experience — discover and unleash their power. I have no doubt that *Facing the Climate Emergency* will help exponentially grow the mass movement demanding the transformational change necessary to avert climate catastrophe.

— **Zack Exley**, co-founder, New Consensus, senior adviser to Bernie 2016, co-author, *Rules for Revolutionaries*

There are many ways forward to saving as much of the world from the climate crisis as we can, while we can. The first step is by moving through the emotional trauma of what is already happening, and toward action. If you want to be a responsible ancestor and ensure a survivable future for all the babies to come, read this book and share it.

— **Pennie Opal Plant**, co-founder, Society of Fearless Grandmothers and Idle No More SF Bay

Somehow this climate crisis manifesto manages to both terrify and inspire, forcing us to confront the sobering facts and galvanize us to save ourselves. It's a dynamic call to action for this pivotal time and the choices we must make. Otherwise, we're eating our children's future.

— **Roberta Baskin**, board chair, Earth's Call Fund

A brand new way of both thinking about the climate emergency and how each of us can fight to stop it. Margaret helps us see that by facing the truth we can liberate ourselves from the paralysis of fear and fight to save the world we love.

— **Trevor Neilson**, founder, Climate Emergency Fund,
co-founder and CEO i(x) investments

Sitting worrying, as the climate emergency unravels, that the world has gone to hell in a hand basket and there is nothing we can do about it is making us sick. Distracting ourselves doesn't work, medicating ourselves or clicking on the occasional petition won't change the outcome, but what will? Go with Margaret Klein Salamon on this journey for your own and for the planet's sake.

— **Christine Milne AO**, Global Greens Ambassador and
former leader, Australian Greens

How is it possible to truly face into the climate devastation unfolding on Earth right now — and come away feeling inspired, engaged, and motivated for action? At once politically engaged, deeply personal, and spiritually resonant, Salamon shares her own intimate journey that led to her leadership of one of the world's most transformational climate organizations.

— **Jeremy Lent**, author, *The Patterning Instinct*

Kids all around the world are on the frontlines of fighting the climate crisis but we cannot do it alone. Margaret exemplifies what it means to be an adult ally and stand up for the youngest generations.

— **Katie Eder**, 19, executive director, Future Coalition,
coordinator of the US Climate Strike Coalition

FACING THE CLIMATE EMERGENCY

HOW TO TRANSFORM YOURSELF WITH CLIMATE TRUTH

MARGARET KLEIN SALAMON

WITH MOLLY GAGE

This book is for all life,

and those who fight to protect it.

Cover design by Katharine Woodman-Maynard and Diane McIntosh.
Cover images: night sky — ©iStock; people with turbine: outline image supplied by Margaret Klein Salamon.

Printed in Canada. First printing April, 2020.

Inquiries regarding requests to reprint all or part of *Facing the Climate Emergency* should be addressed to New Society Publishers at the address below. To order directly from the publishers, please call toll-free (North America) 1-800-567-6772, or order online at www.newsociety.com.

Any other inquiries can be directed by mail to

New Society Publishers
P.O. Box 189, Gabriola Island, BC V0R 1X0, Canada
(250) 247-9737

LIBRARY AND ARCHIVES CANADA CATALOGUING IN PUBLICATION

Title: Facing the climate emergency : how to transform yourself with climate truth / Margaret Klein Salamon with Molly Gage.

Names: Salamon, Margaret Klein, 1986- author. | Gage, Molly, 1978- author.

Description: Includes bibliographical references.

Identifiers: Canadiana (print) 20200159119 | Canadiana (ebook) 20200159259 | ISBN 9780865719415 (softcover) | ISBN 9781550927351 (PDF) | ISBN 9781771423304 (EPUB)

Subjects: LCSH: Environmental degradation—Psychological aspects. | LCSH: Global environmental change—Psychological aspects. | LCSH: Climate change mitigation—Citizen participation. | LCSH: Green movement—Citizen participation.

Classification: LCC GE140 .S25 2020 | DDC 363.7—dc23

New Society Publishers' mission is to publish books that contribute in fundamental ways to building an ecologically sustainable and just society, and to do so with the least possible impact on the environment, in a manner that models this vision.

TABLE OF CONTENTS

PROLOGUE

DID YOU KNOW YOU HAVE A CALLING?

An epic calling.

A heroic calling.

It's probably grander than anything you had let yourself imagine, outside of your dreams. You are *supposed* to save the world. That's why you are here, alive in this time of great consequence.

We — humanity — are putting together a team of heroes to cancel the apocalypse, to protect ourselves and the natural world from catastrophic collapse.

You might not realize it, but you are on the roster. Your jersey is sitting in your locker. We need to figure out your position and get you into (emotional) shape. The first step is to show up to practice. We are waiting for you.

Unlike most self-help guides, my goal is not to make you happy, and it's certainly not to help you avoid pain. This is not about feeling good or finding satisfaction — though these will likely be side effects of fully embracing your mission and living in climate truth.

My goal is to help you maximize your potential to meet the greatest challenge humanity has ever faced. I will show you how to face your pain with courage and vulnerability, and let it motivate you to become the most effective climate warrior you can be.

There is no time to waste. Let's get started.

INTRODUCTION

WE ARE IN A MOMENT OF ACUTE COLLECTIVE SUFFERING. Suicides are up — at their highest point in 50 years — and are now the second leading cause of death for Americans under age 35.[1] One in six Americans takes psychiatric medication, primarily for depression and anxiety. [2] Opioids kill more Americans than car crashes.[3] Virtually all of us resort to something for numbing and distraction: We watch 33 hours of TV a week, scroll endlessly on social media, play video games, and watch pornography.[4] We drink too much, eat too much, work too much, compete too much, and buy too much. Simply put, Americans — and people all over the world — are in pain.

There is, of course, an enormous self-help industry dedicated to helping us feel better. Books, podcasts, and seminars say we're unhappy because we are disconnected from other people and overconnected to technology; they say that we are harshly self-critical and care too much about what other people think, that we live in a suspended adolescence, that we don't manage our money properly and don't take enough emotional risks, that we aren't living our best lives, and that we don't practice enough self-care. Others rightfully point to the glaring realities of inequality, poverty, and precarity as the cause of our pain. These narratives are each important and true to some degree, and they may offer some help. But there's something else going on; something is eating at us. We are in pain because our world is dying and, through our passivity, we are responsible for killing it.

This may sound strange at first. We are so profoundly dissociated from the natural world that we may not register its destruction as something that can profoundly affect our mood. However, think about what happens to you when you see wildfires consuming the Amazon, Australia, and California, floodwaters drowning farmland in France and the Midwest, and hurricanes decimating Caribbean islands. How do you feel when you notice the changing climate in your city? Do you feel terror, grief, anger, and dread? Or is it vague, gnawing anxiety? Do you feel overwhelmed? Or do you feel nothing at all — are you numb to the seemingly impending catastrophe?

Inside all of us, a battle rages. It's the battle between knowing and not knowing, between fully facing the truth — emotionally, as well as intellectually — and shrinking from it. We sense we're in a climate emergency and mass extinction event, but we have a deep-seated psychological instinct to defend against that knowledge. The pain is shouting at us: "Everything is dying!" Somewhere inside, we know that humanity and the natural world are in peril. Indeed, we *feel* the horrors of civilizational collapse and the sixth mass extinction of species, in our bodies. Our minds attempt to shield us from this pain — we avoid, distract, deny, and numb ourselves. But these defenses work only temporarily: When we fail to process our emotions and mourn our losses, the pain takes on tremendous power. It follows us around like a shadow, and we become increasingly desperate to avoid what we know.

This pain has several dimensions. It is the fear we feel for ourselves, for our loved ones, and for all humanity; it is the empathy and grief we feel for the people and species already immiserated or killed; it is the crushing guilt that we feel for continuing to let this happen. Our pain is the consequence of our participation in a destructive system. We have allowed ourselves to become killers — a plague on the rest of life. We share, to varying degrees, guilt and responsibility.[5] Our pain may feel terrible, but it is rational, appropriate, and deserved. It is an internal reflection of external reality: The biosphere — all life — is suffering or threatened. Of course we feel sad and anxious. We are caught in an economic and political system that fosters our collective participation in our planet's daily degradation. Why would we expect to feel good, and good about ourselves, while we are a part of killing all life on Earth, including ourselves and everyone we love?

On one hand, we are victims. No one asked to be born into this broken system that treats human and other life as disposable, allows for unprecedented levels of inequality, and has ignored the climate and ecological crises for decades. We have been failed by the people and institutions that were tasked with protecting us — first and foremost our governments and elected representatives. Our government's failure could not be more complete. The total abdication of duty to protect humanity and all life has made the social contract between government and citizens a sick joke. But the government is not alone: Media outlets, universities, churches, museums, labor unions, environmental organizations, professional associations, and countless others have also failed to acknowledge and protect us from the climate emergency.

And, of course, corporations, such as fossil-fuel companies like ExxonMobil (and their executives) bear an enormous amount of responsibility for the coming global cataclysm. For decades, the fossil-fuel industry has run a multibillion-dollar campaign of lies and climate denial, and it has successfully sowed doubt in our society and blocked anything approaching an appropriate response from our elected leaders. The level of cravenness required to lie to the public about catastrophic warming to continue our addiction to fossil fuels is appalling, criminal, and terribly dangerous.

Many other corporations are also implicated. Big agriculture (particularly Monsanto), big banks, airlines, carmakers, and others have pursued a similarly environmentally devastating business model — contributing to a coming mass death in exchange for short-term profits.

But we are not *merely* victims. Through our participation in this system, through our passivity, we are also perpetrators. We have failed ourselves and each other. We've plundered our home without consideration or restraint, and now we are watching it burn. Although humanity has become almost godlike in our power to create and destroy, we have remained childlike in our use of that power.

In 1956, psychoanalyst and antinuclear activist Erich Fromm wrote *The Art of Loving*, which examined the psychological impacts of a consumer–capitalist society on individuals. Fromm argued that people are alienated from their work, from themselves, and from each other. Fromm noted that people had been sold the view that life was one big competition or marketplace and that people were commodities who should try to maximize not just their money but also their popularity and attractiveness. He observed that people in these societies treat themselves like commodities in a competitive market, adopting false selves to fit in and be liked, while abandoning their authenticity and sense of true purpose.

This ideology still prevails today and fosters the following beliefs:

- You are an isolated individual, defined by what you achieve and what you buy.
- You should focus on competition with others and personal indulgence.
- The only way to have love and acceptance is to own more things.
- There is no community, and there is no web of life.
- Other people are threatening, especially people who are a different race or are from a different culture.
- You have no moral responsibilities. In fact, you are a deprived victim who deserves much more than you get.
- You are living at the pinnacle of human achievement, defined by constant economic growth, and it's naïve to think there could be anything different.

- You may be feeling unpleasant feelings, but they will go away if you buy something.

In his essay, "Love of Death and Love of Life," Fromm postulated that the only reason people would not rise against the possibility of worldwide nuclear destruction was that they were *already* experiencing devastating destruction, internally. On some level, Fromm reasoned, the destruction must have felt appropriate and even appealing — better, at least, than a bullshit, dead-end, alienated, and humiliating life. Otherwise, why did our society allow the risk of mass nuclear obliteration to threaten us for decades? Fromm believed that if people inherently felt their lives were precious and worth living, if people felt engaged in life and saw that engagement reflected in others, if people were not housing a deadness within, they would *demand* an end to the creation of weapons of mass destruction. They would refuse to accept the possibility of the end of all life.[6]

When it comes to the climate crisis, we must ask ourselves the same question: When faced with our current and coming ecological disaster, why are we passively accepting collective suicide and the mass murder of all life? Have we come to feel the pull of death more powerfully than the pull of life?

When we see the media address dire scientific reports in a few stern sentences before cutting away to celebrity gossip; when we see passivity and resignation to our fate from friends and community members; when we hear the common refrain of "We're fucked"; we have to conclude that the coming ecological crisis must feel like an *expected* and maybe even an *appropriate* end to our obviously degraded society. How are we otherwise able to make sense of the fact that people aren't rioting in the streets at the imminent destruction of their lives, their children's lives, and the entire web of life? Do we want to live? If we do, we need to wake up and grow up — right now. We are about to lose everything, but we aren't dead yet.

Fromm describes the pull of death as "necrophilia." This concept is informed by Freud's theory of the "death instinct." It isn't a sexual dynamic — it's a deeply personal orientation. Necrophilia is a love "for all that is violence and destruction; the desire to kill; the worship of force; attraction to death, to sadism; the desire to transform the organic into the inorganic by means of 'order.'"[7] Necrophilia prioritizes things over people, and possessions over experiences. "The necrophile," writes Fromm, "lacking the necessary qualities to create, in his impotence finds it easy to destroy because for him it serves only one quality: force."[8]

Our society treats life — human life, plant life, animal life — as if it were a cheap commodity rather than the most precious, sacred thing there is. By doing so, we've not only ensured the coming ecological crisis; we've inured ourselves to it. It seems to us to be an appropriate end: Worldwide annihilation reflects the emotional and spiritual annihilation we've internalized.

It doesn't have to be this way. In fact, we can face climate truth and choose not to commit passive suicide.

We can choose to turn away from illusion and distraction. We can each decide to face climate truth and make the choice that *now* is the time to do everything in our power to wrest life back from the jaws of extinction. We can each help to initiate a collective awakening to the climate emergency and a World War II-scale response that protects humanity and the natural world and builds a beloved community.

To do so, we must shake off our resignation and our narcissistic, consumerist, necrophilic orientation, and denounce the lies that aid and abet denial. Each of us must do our part to reestablish our connection to all life and to recognize our bottomless responsibility to protect it. We must acknowledge that responsibility. We must take on the mantle of "protector" or "warrior" with joy and pride. We must join the team to protect humanity and all life. We can allow ourselves to face the truth and to accept the reality that we must transform — *now* — individually and together, to respond effectively to the climate crisis.

Socialism has experienced a resurgence in recent years — partly because many people see capitalism as squarely to blame for the climate emergency. It's true: Capitalism, with its dependence on endless growth, its staggering levels of inequality, its treatment of workers as disposable and the living world as expendable, and its relentless use of advertising to make good citizens synonymous with good consumers, is a key part of the problem.[9]

However, it's simplistic to blame capitalism alone. Governments and citizens worldwide have perpetrated the damage to our planet, and it hasn't taken place solely under the auspices of capitalism. The communist, totalitarian Soviet Union was approximately as damaging to ecosystems as the market-based United States[10] and was the second-largest emitter of greenhouse gases during the 1960s, '70s, and '80s.[11] China, with its mostly state-driven economy, has now become the world's largest emitter of greenhouse gases. The social democratic state of Norway owns 67 percent of Equinor, formerly Statoil — an oil and energy company.[12] The change we need to make is *even bigger* than shifting to socialism.

To solve the climate and ecological emergency, we must transform our destructive economy into a *regenerative* one, and we must do it at emergency speed. We don't just need zero emissions in every sector; we need huge carbon drawdown projects that restore ecosystems and the soil. We need permaculture and food localization; we need an end to mass consumerism and endless growth; we need to give back half the Earth to nature to restore biodiversity;[13] and we need to create a society based on protecting and healing humanity and the natural world. This means transforming not only our energy, agricultural, transportation, and industrial systems — it means transforming ourselves and how we relate to each other. We need to rethink our basic concept of who we are and what matters.

And we need to do it all *right now*.

This is a self-help book, but its goal is not to make you feel less pain. Its goal is to make you feel your pain more directly and constructively: to turn it into action that protects humanity and all life. In this book, I argue that your *pain is a signal* — it's telling you something critically important. The pain is demanding to be acknowledged, and I want to show you how to listen to it. I want you to face the pain of the climate and ecological emergency, and to feel it in a focused, conscious way so that you can initiate a process of transformation — first in yourself and then in society as a whole. This large-scale change must be our goal, as Pope Francis wrote in his 2015 encyclical *Laudato Sí*. To stop the climate emergency, he says, we must "become painfully aware, to dare to turn what is happening to the world into our own personal suffering and thus to discover what each of us can do about it."[14]

We can use our pain to effect tremendous change. I know because I've been through this process myself: I've felt the pain, I've faced it, and I've used it to motivate myself and others for change. Before I started a climate advocacy organization, I was a young professional in New York City — a clinical psychologist working on a doctoral degree, preparing to enter private practice and start paying off my six-figure student debt. I wasn't a climate denier, but I was willfully ignorant and disengaged. I avoided thinking or reading much about the climate because it made me feel terrified and helpless. I would read the first sentences of articles about global warming, say to myself, "Nope! I can't do it," close the article, and distract myself with something else.

But in October 2012, Hurricane Sandy hit, and New York City came to a standstill. Destruction was everywhere. I vividly remember seeing a car that had been smashed by a huge branch.

It had a cardboard sign on the shattered windshield that read, "Is global warming the culprit?" Seeing the message caused something in me to shift. I *knew* the answer to that question, though my knowledge was diffuse, even unformulated. But that sign helped me bring my awareness into stark focus: If global warming had smashed that car and the whole city, what else could it do? How bad was this situation, and what did our collective future hold? With these questions in mind, I started to educate myself. I began to finish the articles that had previously overwhelmed me. I started to seek out books on the climate and ecological emergencies.

What I learned shook me to my core — and caused me to reassess my life. I realized that it was my responsibility to do everything I could to halt and reverse the coming catastrophe. So I left the field of psychotherapy — which I loved — to found and direct The Climate Mobilization (TCM), an organization that tells the truth about the climate emergency and advocates a WWII-scale transformation of our economy and society to protect humanity and the natural world. To give TCM the best chance of success, my husband and I focused on necessities, moving into a small, affordable apartment so that I could build the organization as a volunteer.

Through my work at TCM, I have had hundreds of conversations with people — as diverse as elected leaders, climate scientists, stay-at-home moms, hedge-fund managers, and janitors — about how to process fear and respond to climate truth. I've facilitated discussions about "life in the climate crisis" and how to turn "pain into action" for all kinds of groups, whether in person, on phone calls, or through Facebook. Through my work, I have seen individuals perform extraordinary feats of service — people rearranging their lives, leaving their jobs, spending their savings, moving in with their parents — to go "all in" for this mission.

I have also seen a powerful movement arise. I have seen Extinction Rebellion (XR) use nonviolent, direct-action strategies to declare that the government has fundamentally broken its social contract and that rebellion is our best hope of survival. I have seen students walk out of their classes *by the millions*, demanding that adults take responsibility for solving this emergency. I have seen the Sunrise Movement mobilize for a Green New Deal — a WWII-scale response to the climate emergency. I have seen the beginning of a collective awakening, with more and more people deciding that they will not sit quietly while the world around us dies.

There is a movement being born, and I am proud to say that I helped bring it into existence. The members of this movement are not content to numb our sadness with money and things. We're not willing to ignore the Earth as it burns. We're going to fight for what matters. We know that we can face climate truth and let it transform us.

This book will show you how to join our ranks as members of the climate emergency movement. In it, I will ask you to tap into your fear about our current climate crisis and the future we are careening toward. I will help you mourn what has already been lost and what we continue to lose every day. I will help you transform your despair into a collective effort to build power for the movement.

It's not going to be easy. It's going to be the opposite of easy. But acknowledging the truth of our climate and ecological emergency, grieving the past and the future that has been lost, and taking the heroic path of the climate warrior will make you confident and strong. It will give you a mission and purpose beyond anything you have experienced before. It will allow you to, at long last, heal your pain and feel genuinely good about yourself. It will connect you to your fellow humans, and it will connect you to all life. It will give you real, meaningful hope because it is based on your real and incredible potential to affect change. Most importantly, it will help give humanity a better chance of canceling the apocalypse and protecting itself and the living world.

It will be a difficult journey, but I can promise you that when you commit to taking it, something wonderful will happen. You will feel hope; you will know that you are part of the solution; you will know that you are doing your part to save the world. You will become a climate warrior, leveraging your strengths and your capacities and mobilizing to save the world. You *can* transform yourself with climate truth and become the hero humanity needs you to be. No one is coming to save us, but together we might be able to save ourselves.

Questions for Reflection and Discussion

- Do you feel anxiety, depression, or have other painful psychological experiences that you struggle to define?
- How might some of those feelings be driven by the climate crisis and species extinction?
- Have you experienced any sense of the inner-deadness that Fromm describes?
- How and to what extent have you replaced your love of life with a love of objects?
- What do you most love about humanity and the living world? What do you most want to protect?
- Can you imagine yourself as a hero? As a protector? As a climate warrior?

STEP ONE:
Face Climate Truth

YOU CAN CREATE TRANSFORMATIVE CHANGE ONLY by facing the truth. That's why the first step toward transformation is to acknowledge climate truth with your whole self — with both your intellect and your emotions. What do I mean by "climate truth"? Simply this: Scientific consensus says a climate emergency and an ecological crisis threaten everyone on this planet. Only an emergency mobilization of resources to rapidly transform our entire economy and society can protect us.

This might feel like the hardest part. The reality of the current and coming catastrophe can feel too big and too overwhelming, so many of us turn away, telling ourselves that we can't handle it, that we're just not ready. This is an understandable response — I've had it myself — but it will not protect humanity or the natural world. Furthermore, avoiding the truth requires constant effort and vigilance. You may not even realize how much energy must be spent on *not* acknowledging environmental collapse. It is hard work not letting yourself feel your fears. When you avoid the truth, you put the energy that could be used toward *preventing* the climate emergency into safeguarding the fiction you've created for yourself.

The questions of whether or not you're "ready" to face the reality of our climate emergency or whether or not you can "handle it" are the wrong questions. To begin acknowledging climate truth, ask yourself, *"What is my priority?"* Would you rather protect yourself from painful knowledge or protect yourself, your family, and the entire human family from the actual climate emergency? These are our two options. The choice should be easy.

A truth, even a painful truth, is powerful. Have you noticed how good, how freeing, how fortifying it feels to tell the truth?

Have you seen how, in your personal life, it has been a crucible for growth? When you face the truth with your mind and heart, you will experience this same freedom. When you face the truth, you will allow yourself to move through anger and grief to hope and action. It's hard at first; it can even feel impossible. But it is an incredible relief to let go of your defenses, your vigilance, your effort to safeguard and deny. It is an incredible relief to allow yourself to live in truth. It may be hard to believe, but fully integrating climate truth into your life will make you lighter, less encumbered, and more capable of facilitating change.

Facing climate truth means recognizing, as journalist David Wallace-Wells writes in the opening lines of *The Uninhabitable Earth,* that "it's worse, much worse, than you think."[15] It means realizing that many contributors are creating an unrealistically optimistic picture of the climate emergency. These contributors include the fossil-fuel industry's misinformation campaign, the failure of both political parties to reckon honestly with the emergency, the corporate media's shameful silence, the Intergovernmental Panel on Climate Change's (IPCC) systematic bias toward understatement, the misguided euphemisms of the gradualist climate movement, and our personal defenses.

Thanks to American scientist Eunice Foote and Irish physicist John Tyndall, we've known since the 1850s that carbon dioxide (CO_2) is a heat-trapping gas.[16] In 1898, Swedish scientist Svante Arrhenius identified the global greenhouse effect, theorizing that burning fossil fuels could cause the atmosphere to warm.[17] Additionally, English engineer Guy Callendar demonstrated in 1938 that the Earth was already warming due to carbon dioxide emissions, though it took a couple of decades for his work to gain scientific acceptance.

The facts about CO_2 and the greenhouse effect were not limited to academics. As has been well documented, ExxonMobil's scientists have been aware of the greenhouse effect since the 1970s.[18] For some time, they even had a team of top scientists conducting pioneering research into the problem. But rather than using that information to inform the public of the existential risk and transitioning their business so that it would not destroy civilization and the living world, ExxonMobil launched a massive disinformation campaign to create doubt about the scientific certainty of the greenhouse effect. To extend the campaign's reach, Exxon and other fossil-fuel companies funded denialist scientists like Wei-Hock "Willie" Soon and denialist think tanks like The Heartland Institute (THI).[19] THI's campaign included, to give one example, mailing its book and DVD, *Why Scientists Disagree About Global Warming*, to 350,000 American public-school science teachers in 2017.[20]

This campaign of lies and political manipulation has cost fossil-fuel companies hundreds of millions of dollars, and it has been remarkably successful.[21] If you have ever had any questions or doubts about whether the greenhouse effect is a scientific consensus, it's because the oil industry's campaign has worked. If you have ever allowed yourself not to worry about the climate emergency because "scientists are still figuring it out," it's because of the skepticism the industry has cast on basic science. If you have ever held yourself back from discussing the climate emergency because "it's too controversial," it's because the oil industry has strategically invested in your silence.

The denial campaign has been successful because it aligns with our desires and defenses. The oil industry is telling a story we all wish was true. Most of us would prefer not to have to face the climate emergency and the "inconvenient truth" that our primary power source for electricity, transportation, and manufacturing is deadly. We would rather not think about it, not have an emotional reaction to it, not talk about it, and not recognize solving it as our responsibility. We use every psychological defense:

- Denial: *It's not real.*
- Intellectualization: *It's real, but it doesn't affect me emotionally.*

- Willful ignorance: *I don't want to know what's going on; it's too scary.*
- Wishful thinking: *It can't be that bad.*
- Regression: *We need the experts to handle this.*
- Rationalization: *I can't do anything meaningful to address it.*
- Compartmentalization: *It's not relevant to my feeling like a good, moral person.*
- Projection: *It's happening, but it's other people's fault and responsibility.*
- Dissociation: *I am going to zone out or numb myself with substances, video games, or other distractions.*

We also share a common tendency to focus on individual consumption choices, in which we might say to ourselves, "This is happening, and it's my responsibility to purify my consumption and reduce my carbon footprint." While this is, in some ways, laudable, it can also be a defensive distraction from creating transformative societal change.

At the same time that the fossil-fuel industries were building the disinformation campaign to feed our defenses, fossil-fuel magnates David and Charles Koch and other fossil-fuel interests were successfully preventing our government from taking the climate emergency seriously. Fossil-fuel companies spent more than $2 billion lobbying against climate legislation between 2000 and 2016.[22] The GOP is the only major political party in the world that continues to deny the science of global warming.[23] And for decades, the Democratic Party has wildly understated the threat and the scale of response required to protect humanity and the living world. Trump is merely an exaggerated embodiment of the firmly entrenched, bipartisan commitment to ignore and underplay the crisis.

Presidential administration after administration has failed to reckon with the scale of the crisis. In the early 1990s, George H.W.

Bush's words at the 1992 Earth Summit — "The American way of life is not up for negotiations" — laid a strong foundation for administrative support of wanton consumerism.[24] Obama boasted about increasing the United States' oil production "every year I was president" and turning the United States into the world's largest exporter.[25] This was in 2018, as the climate was spinning out of control.

The Intergovernmental Panel on Climate Change (IPCC), the United Nations' body driving much of the climate change conversation, has also failed to fully acknowledge and communicate the climate emergency, precisely because it is an inter*governmental* panel and not a purely scientific body. Governments appoint the lead authors of the IPCC's scientific reports, and all members — including petro-states like Saudi Arabia, Russia, and the United States — directly influence the IPCC's *Summary for Policymakers* reports, which drive media coverage on climate change.[26] Climate-policy analyst David Spratt and former fossil-fuel industry executive Ian Dunlop demonstrate how "scholarly reticence" and governmental influence in the IPCC have created a systemic bias toward euphemism and an understatement of existential risk. This understatement actively deepens the public's ignorance, as Spratt and Dunlop argue in their report, *What Lies Beneath*.[27]

Even the climate and environmental movement understates the risks of the climate emergency. Indeed, in 2017, when Wallace-Wells published his wildly popular article on the possible worst-case scenarios of the climate crisis, he was chided from within the climate movement as "doomist." People who spoke about the coming climate catastrophe were reminded, once again, that "fear does not motivate, and appealing to it is often counterproductive as it tends to distance people from the problem, leading them to disengage, doubt and even dismiss it."[28]

These comments reflect what has become orthodoxy in the mainstream, or what I call the "gradualist" climate movement: *We must not scare the public; they cannot handle it.* This misguided and counterproductive mandate has its origins in the culture of science, which tends to treat emotion as a threat to rationality. The "fear of fear" is reinforced by philanthropy, which is funded by corporations and the very rich, who generally prefer cheerful optimism and who more frequently fund reformist political advocacies like carbon pricing, instead of investing in a movement for transformative change.[29]

There are some obvious problems with the implicit decision made by scientists, the climate movement, and the media to avoid scaring people. Principally, it ensures that people are unprepared for a truth that is, in fact, frightening. When it comes to discussing the climate, the widespread fear of fear means that the only option is to *avoid telling the truth*. This is exactly what happens. Very few people, including gradualist climate leaders, talk about the risk of the collapse of civilization or the deaths of billions of people, even though we are careening toward these catastrophes.

This recommendation to offer euphemism rather than fact is made explicit in Columbia University's popular guide, *The Psychology of Climate Change Communications*, which devotes an entire section called "Beware the Overuse of Emotional Appeals." The guide advises communicators to selectively "decide what portfolio of risks they want to make the public more aware of... such as the relationship between climate change and disease."[30] "Emotional appeals" (such as telling the whole truth about the coming collapse of civilization) could make people "numb" or "backfire down the road."[31] In other words: Don't tell the whole truth. Americans are too weak and childlike to handle it.

The claims that "fear doesn't work" are not only patronizing and cynical; they have also been devastating in terms of mounting a real and timely response to the crisis. They are *not* supported by evidence.[32] Further, the hollow optimism and positive messaging have tripped the public's bullshit detector. People can tell when they are being given a canned message rather than the candid truth. We know, with varying degrees of conscious awareness and intellectual understanding, that the Earth's systems are deteriorating — and rapidly.

Again and again, leaders of so-called "big green" organizations, including the Environmental Defense Fund, the National Wildlife Federation, and the National Resources Defense Council, have failed to offer a frank assessment of the climate crisis and have failed to dedicate their more than $100 million in annual budgets toward advocating for solutions that stand a chance of protecting us.[33] Like the Democratic Party, the mainstream climate movement (and the vast majority of civil society institutions) is stuck in the stultifying morass of gradualism with a completely ineffective focus on "business as usual."

Far too many climate organizations have not adopted the 10-year decarbonization timeline — instead, they call for eliminating emissions by 2050. At this late stage, when the crisis is accelerating and tipping points are being reached, this gradualist timeline is irresponsible. And then there are the climate organizations still talking about the need for reducing, rather than eliminating, emissions. This ignores the reality that as long as we emit greenhouse gases into the atmosphere, we will continue warming the planet. We need to be carbon *negative*. Few large, mainstream environmental groups are calling for actions to draw down, or sequester, excess greenhouse gases, although we know that this must begin *now* — and on a massive scale — if we are to avoid a worst-case-scenario climate catastrophe.

These environmental organizations, like so much of our society, have had their political imaginations constrained by neoliberal ideology. They have come to believe that only minor, market-based reformist change is possible. In the neoliberal mind, life is nothing but a competitive capitalist marketplace. Everyone is — and must be — out for themselves. According to this hyper-individualized mindset, there is no shared belief that *we*, the people, make up society; that *we*, the people, can determine an alternate path because *we, the people, have the ultimate authority over our collective fate.*

These factors all add up to a shocking misjudgment of the climate risk. The "Overton window," also known as the window of political discourse, describes the range of ideas discussed in the media and considered publicly "acceptable." Because public discussion is trapped between denial and gradualism (and the IPCC report), there is not yet room in the Overton window to accommodate the reality that global warming is occurring more quickly and more intensely than reported by the IPCC. The graph below, from atmospheric and oceanic scientist Michael Tobis, editor-in-chief of the sustainability website Planet 3.0, illustrates the ways that the climate conversation fails to consider catastrophic scenarios — even though they are likely.[34]

The Earth has already warmed 1.1 degrees Celsius (°C) above late-19th-century temperatures. Even this amount of warming is creating millions of climate refugees and killing 400,000 people a year.[35] Meanwhile, the widely publicized 2015 Paris Agreement on emissions reductions offers a path to a whopping 3.5°C of warming. Such an outcome could bring human civilization as

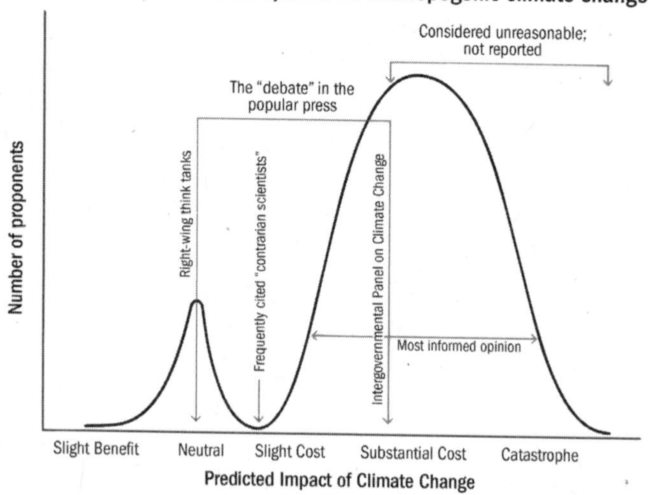

CREDIT: MICHAEL TOBIS AND STEPHEN BAN.

we know it to a violent end.[36] Rather than a gradualist approach championed by governments and various mainstream organizations, a full-scale mobilization of society for zero CO_2 emissions, plus drawdown, is our last, best hope of avoiding apocalyptic catastrophe and restoring a safe and healthy climate. Even the IPCC, with its penchant for understatement, recognizes that:

> Pathways limiting global warming to 1.5°C with no or limited overshoot would require rapid and far-reaching transitions in energy, land, urban and infrastructure (including transport and buildings), and industrial systems (high confidence). These systems' transitions are unprecedented in terms of scale.[37]

Related ecological emergencies are exacerbating the emergency of global warming. Through a combination of ecological crises, including habitat destruction and biodiversity loss, phosphorous and nitrogen flows from fertilizers, and toxic pollution, the Stockholm Resilience Centre argues, we have overstepped the boundaries necessary for a "safe operating space for humanity."[38]

Extinctions are happening at about 1,000 times the rate that we would expect without human influence, with dozens of species going extinct every day.[39] The World Wildlife Fund's *Living Planet Report 2018* documents a 60 percent decline in the populations of mammals, fish, birds, reptiles, and amphibians in the past 40 years.[40] The "insect apocalypse" is the tip of the spear, but it's also the base of the food chain. A German study found that the number of insects, surveyed by weight, has decreased by 75 percent between 1989 and 2016.[41] In North America, the Monarch butterfly population has crashed, declining 90 percent in the past 20 years.[42]

The fact is, as Australian environmentalist and former CEO of Greenpeace Paul Gilding says in *The Great Disruption*, "The Earth is full. In fact, our human society and economy is now so large we have passed the limits of our planet's capacity to support us and it is overflowing... [I]f we don't transform our society and economy, we risk social and economic collapse and the descent into chaos."[43] In other words, the global economy and human population, and the domesticated animals we raise and consume as food and products, have grown so vast and unsustainable they undermine the life-support systems required for our collective survival. According to the Global Footprint Network, humanity is using the equivalent of 1.7 Earths per year.[44] Back in 1970, they calculated that we were using the equivalent of one Earth per year.[45]

On our current trajectory, we're facing civilization's collapse. Our society — our financial, cultural, educational, governmental, and media systems — was built on a safe and stable climate. Altering that foundation is causing an array of dire impacts, from unlivable heat stress to uncontrollable wildfires and mudslides, from super-hurricanes and multi-meter sea level rise to pandemics and water wars. Environmental analyst Lester Brown explains that the climate impacts most immediately threatening civilization are droughts, which, combined with topsoil loss, lead to crop failure, which prompts forced migration, which causes social destabilization, which provokes state failure.[46] It's already happening: The worst drought in Syrian history caused massive internal migration and unrest, setting the stage for a brutal civil war.[47]

In 2008, Canadian journalist Gwynne Dyer wrote *Climate Wars,* which considers possible scenarios faced by humanity as the world warms. According to Dyer, after a warming of 2.8°C, the geopolitical state of the world in 2045 could be marked by the breakup of the European Union (EU) and the closed borders of northern European nations, the governments of which would seek to keep out refugees from "famine-stricken Mediterranean countries." Dyer predicted that a wall would be built on the United States' southern border, more countries would acquire nuclear weapons, and tens of millions of people would die from starvation. In one scenario, he predicted that countries like Britain and Japan would barely manage to feed themselves and would close themselves off, and millions would die from starvation and anarchy in more vulnerable countries like Saudi Arabia and Uganda. While it may have seemed like dystopian fiction rather than a brilliant prediction, in the decade since the book's publication, Dyer's scenarios have begun to unfold. Britain is leaving the EU as migration pressures and the population's xenophobic reaction to immigrants mount. The southern Mediterranean area is drying.[48] Climate is a driver of state instability across the Sahel.[49] And, of course, the United States elected a president whose most prominent campaign promise was to build a wall on the southern border.

Dyer's scenarios demonstrate how Brown's analysis could play out — how civilization could collapse after droughts that cause food shortages, which cause failed states — and its plausibility is incredibly alarming. Ongoing catastrophes and mass migration will require vast resources, tax social welfare and international aid bureaucracies, weaken family and community support structures, send tremors along existing cultural fault lines, and whip up resentment and racism. The privileged will attempt to protect themselves and let the less privileged starve, but even this level of selfishness and inhumanity will not succeed for long. No one will be protected in the wake of civilization's collapse.

I know these scenarios are terrifying. I am intimately familiar with the appeals of defending against and blocking them out. I've already described how, for many years, I did just that, living in the uneasy peace of denial. In high school in Michigan and college in Massachusetts, I was willfully ignorant of global warming and other ecological threats. I was always more interested in people than in nature, more disposed to the humanities and social sciences than to the hard sciences. I found discussions of environmentalism and climate change boring. I thought, "Why am I supposed to care about melting ice, polar bears, or two degrees of warming?" It just didn't seem to add up to very much.

I see now that I didn't *want* it to add up to very much. I didn't want to consider that something so seemingly unconnected to me, yet so apparently huge, was something I needed to pay attention to, something that defined my future. So for years, I utilized the defense of willful ignorance. I've already described how I'd start to read an online article about the climate crisis but get overwhelmed by fear, close the tab, and tell myself, "I just can't handle this." A part of me already knew enough to realize I didn't want to know more.

But over time, as I became more mature and lessened my defenses through psychotherapy, I became increasingly worried about the changing climate. A few events contributed to my transformation, which was nearly complete by the time Hurricanes Irene and Sandy hit. In my senior year at Harvard, I took a course on agriculture to fulfill my science requirement. It opened my eyes to how at-risk the global food system is to climate and ecological threats. After graduating, I backpacked in Europe and Brazil. During my journey, I met many young people from Europe and Latin America who were putting sustainability at the center of their career plans. I also saw that many European cities were leagues ahead of the United States in terms of mass transit and walkability.

It still took me years to fully reckon with the climate emergency. When Hurricane Sandy hit while I was living in New York City, that foreboding sign — "Is global warming the culprit?" — prompted me to discuss my fear of the climate crisis in therapy. My therapist, who thought I was overreacting, told me, "You know, you worry a lot about the climate, but you don't seem to know much." I found this to be a provocative and galvanizing statement, partly because she was right. I had been willfully ignorant for so long that my understanding of the climate crisis was impressionistic and vague.

After my therapist's challenge, I started to finish the online articles. Then I started reading books and examining scientific studies on the climate and ecological emergencies. The truths I read were shattering. I still remember the force of Bill McKibben's argument in his book, *Eaarth*: We have made such fundamental changes to our planet that the Earth needs a new name.[50] He opens the book with this haunting image:

> Imagine we live on a planet. Not our cozy, taken-for-granted Earth, but a planet, a real one, with melting poles and dying forests and a heaving, corrosive sea, raked by winds, strafed by storms, scorched by heat.[51]

McKibben shows us that we have almost squandered an incredibly rare and precious gift — the Holocene conditions in which civilization developed — and he urges his readers to engage with reality. Reading his book, and others like it, filled me with terror and heartbreaking despair. I felt, as McKibben advises, as though the alarming statistics associated with the climate emergency were "body blows...mortar barrages...sickening thuds."[52]

This visceral pain, more than anything else, forced me to take seriously the fact that our civilization and economy is entirely dependent on our atmosphere and biosphere. We are so dependent on the atmosphere and biosphere that they might as well be parts of our bodies. When I started to think like this, to feel the pain of the climate crisis in my own body, I knew my life had to change. If I continued living focused on myself, I was not only shirking my duty as a protector of humanity and all life; I was putting myself at risk. I had to do everything I could to transform our fossil-fueled society and economy.

Although the climate emergency is terrifying, facing and telling the truth is an incredibly powerful force for change. As you embark on the work ahead, remember that humans evaluate danger and risk by noticing how other people respond. When we see people in our community acting as though nothing is wrong, it is a cue to us and to everyone else that everything is normal. When we see people in our communities responding to an event as though it is an emergency, we start to view the event as an emergency, too. Telling the truth about the climate, and treating the climate crisis like the emergency it is, is *highly contagious.*[53]

To put this in more relatable terms: Imagine you are at work, and you hear the fire alarm go off in your office building. You don't see a fire and you barely smell a whiff of smoke. How seriously should you take a fire alarm if you aren't sure you're in danger? What if it's just a drill? What if a coworker overheated something in the microwave again and triggered the smoke detectors?

When you hear an emergency signal but you aren't sure if the emergency affects you, you seek answers to questions like these by watching the actions and communications of the people around you, particularly those people designated as leaders.[54] If the fire alarm goes off and leaders are chatting and staying put, or leisurely exiting the building, you will likely assume that the fire isn't a real danger or that the alarm is just a drill. When people, particularly leaders, react passively to an emergency, so, too, will nearly everyone else. This phenomenon, also known as pluralistic ignorance, has been proven in psychology labs many times.[55,56]

If, however, when the fire alarm goes off, you see your bosses moving quickly, their faces stern and focused, communicating with each other and with you and your coworkers with urgency and gravity, you will usually assume that there is a real danger and exit the building as quickly as possible.

Emergency mode — and what entering emergency mode means for individuals and societies — is discussed in Step Four. Here, I just want to point out that Americans have entered emergency mode to solve unprecedented emergencies before. Most famously was the Allies entering emergency mode when we mobilized to win World War II. Faced with the prospect of annihilation, Americans and Britons were expected to pull together by working in war jobs, growing victory gardens, contributing to scrap drives, and — in Britain — volunteering as Air Raid Precautions wardens.[57] The U.S. government made an extraordinary investment in the war effort, spending more than 40 percent of its gross domestic product on the war effort,[58] issuing strong regulations, and rationing critical goods like gasoline and meat.[59]

Today, even though it's our entire planet that's burning, there hasn't yet been anything like this emergency response. Our collective denial, fueled by our aversion to feeling fear and grief, holds us back. By denying the truth of the climate emergency, we remain in the thrall of helplessness, our political system remains intractable, and our politicians and mainstream climate movement organizations remain unwilling or unable to lead the way. Virtually everyone — even powerful elites — appear overwhelmed and incapable of processing the immense scale of the climate crisis. We all feel helpless, unable to grasp how we can meaningfully contribute to transforming our world and protecting humanity. We look to our leaders for our cues, but we only see the same passivity, powerlessness, and helplessness reflected back at us.

This powerlessness is a myth rooted in a lack of understanding of how malleable the human condition is; how capable of growth we are; and how people have formed social movements and created fundamental changes in society over and over again. The feeling of powerlessness has been fostered by neoliberal ideology. Instead of viewing ourselves as citizens in a democracy — as people who are responsible for creating the world we want to live in together — we have been taught to view ourselves as a group of self-interested individuals whose only role is to compete in the free market — and "get ours." This ideology has dominated the American and global political imagination for decades. It tells us that our society is a meritocracy that rewards hard work and virtue; it tells us that poverty is a moral failing; it tells us that there is no alternative to our current economic model and that only "market-based" remedies for the climate are "realistic." We've been told, most famously by former British Prime Minister Margaret Thatcher, that "there is no such thing as society," certainly no group of people joined together in the common interest of a habitable planet.[60]

However, I'm here to remind you that groups of concerned citizens have changed the world many times before — and they have done it through the power of truth. Telling the truth, and spreading it rapidly and completely, has been a central strategy in successful social movements. Importantly, such movements do not *uncover* hidden truths. Instead, they make manifest truths that are often already widely known before the movement's emergence but that have been repressed, denied, and ignored by governmental, corporate, media, academic, and religious institutions.

We've seen this dynamic most recently in the #MeToo movement, but it has happened before. Before the Civil Rights movement brought the ugly truth of racism to the forefront of American life, most white Americans ignored or passively accepted the oppressive Jim Crow laws, telling themselves that it was not their problem. Yet, when nonviolent protesters were met with hateful violence and when those events were broadcast on televisions across America, the truth could no longer be denied or defended against. In seeing the clear truth of racial segregation's moral bankruptcy, viewers across America could no longer claim that this didn't concern them. People were forced to choose a side; when more Americans began to take the side of civil rights, seeing it as *their* issue, major, immediate changes were seen as undeniably necessary. When a powerful truth, such as the brutal treatment of African Americans, is effectively communicated, transformation can happen rapidly.

The fall of the Soviet Union offers the context for another example. There, Czech artist and dissident Václav Havel led a successful social movement dedicated to spreading the truth. Havel, who championed "living in truth" rather than complying with the corrupt, repressive actions of the Soviet state, argued that the government's lies would collapse when the force of the truth was unleashed. He was right.

Although the desire for change in Czechoslovakia had been building throughout the 1970s and '80s, dissent was illegal. Most people lived in an enforced state of pluralistic ignorance, in which everyone assumed that everyone else supported the state. Dissidents like Havel worked to contradict and expose this pluralistic ignorance. The illusion was shattered on November 20, 1989, when 200,000 people demonstrated in Prague. When others realized how much power was on the people's side, the number of demonstrators swelled. The next day, the demonstration grew to 500,000 people, and a general strike was called. Ultimately, the entire leadership of the Communist Party of Czechoslovakia resigned; the first democratic elections in 43 years were held; and Havel was elected president.[61] This collective upheaval was possible because, as Havel asserted, the truth has inherent power:

> [The power of truth] does not reside in the strength of definable political or social groups, but chiefly in a potential, which is hidden throughout the whole of society, including the official power structures of that society. Therefore this power does not rely on soldiers of its own, but on soldiers of the enemy as it were — that is to say, on everyone who is living within the lie and who may be struck at any moment (in theory, at least) by the force of truth (or who, out of an instinctive desire to protect their position, may at least adapt to that force). It is a bacteriological weapon, so to speak, utilized when conditions are ripe by a single civilian to disarm an

> entire division.... This, too, is why the regime prosecutes, almost as a reflex action, preventatively, even modest attempts to live in truth.[62]

Although the truth is a radical and incredibly motivating force, it must be made manifest to have power. When it comes to the climate crisis, it's not that the people are powerless, it's that larger forces have conspired, both purposefully and inadvertently, to deny the truth, making us *feel* powerless. Most of what we've been told about the climate emergency is either outright deceit (on the part of the fossil-fuel industry and the GOP) or euphemistic understatement (on the part of the Democratic Party and the gradualist climate movement). Is our inability to effectively respond really surprising? Is it any wonder that so many of us do not — or are simply unable to — acknowledge and respond to the climate crisis as the emergency that it so clearly is?

Thankfully, an alternative to the gradualist climate movement is emerging: the climate *emergency* movement. This movement demands what science and morality tell us are necessary — a 10-year transition to zero emissions plus drawdown. It's led by Alexandria Ocasio-Cortez and the Justice Democrats in Congress; the Sunrise Movement, Zero Hour, school-aged climate strikers, and the Extinction Rebellion in the streets; and TCM in developing and pioneering new campaigns and policy approaches. The climate emergency movement is dedicated to harnessing the truth's radically transformative power to speak what we all sense and to advocate for a policy approach — WWII-scale climate mobilization — that actually has a chance of protecting humanity and restoring a safe climate.

Greta Thunberg, a Swedish teenager who, at 15, started a school strike for the climate, embodies the candor of the new movement. At the 2018 UN Climate Change Conference in Katowice, Poland, she told the diplomats of the world, "You are not mature enough to tell the truth." In 2019, at the World Economic Forum in Davos, Switzerland, inspired in part by the work of TCM and other organizations, she said,

> I don't want you to be hopeful. I want you to panic. I want you to feel the fear I feel every day. And then I want you to act. I want you to act as you would in a crisis. I want you to act as if our house is on fire. Because it is.[63]

It's time for all of us to reckon with that truth. Our common home, the only one we have, is on fire. Can we overcome our fear enough to face reality? Will we wake up and do everything that we can to put out this fire?

Questions for Reflection and Discussion

- Have you found facing the truth to be transformative in other areas of your life? When have you experienced personal growth by facing a hard truth?
- What defenses have you used to protect yourself from climate truth? Do you think you will be able to lessen or let go of those defenses?
- How well do you understand the scale of the climate and ecological crises? Do you feel you need to learn more to engage effectively?

STEP TWO:

Welcome Fear, Grief, and Other Painful Feelings

FACING THE CLIMATE EMERGENCY IS *HARD*. It challenges us to reach for new levels of emotional strength, maturity, and wisdom. That's why a critical part of facing the truth means improving how you relate to your feelings, particularly the difficult feelings brought forth by the climate emergency. Rather than block out or deny these feelings, you must face and work through your fear, grief, anger, guilt, and all the other painful feelings you've repressed. In fact, you must go further than just facing them — you must learn to welcome them. Only then will you escape the control these feelings have over you. Only then will you shake off their numbing and paralytic effect and be able to use their power to transform yourself and our society. In this chapter, I will help you notice your pain with nonjudgmental interest; I will help you allow yourself to feel and even welcome your pain because this pain will propel you to fulfill your potential as a climate warrior.

First, you must learn to feel your feelings. This sounds like easy, straightforward advice, but it's one of the hardest things for humans to do, especially in an alienated society like ours. Growing up, we are typically taught that some feelings are not acceptable, and must be repressed and denied at all costs. For most people, the feelings that were scorned, feared, and otherwise rejected in our families of origin are feelings that we've built a lifetime of defenses against. Often, these lessons take gendered lines: Boys are taught that sadness, fear, and vulnerability are pathetic and embarrassing, while girls are taught that anger and envy make them "bad girls."

Too often, when we look inside ourselves and find "negative" feelings, we judge and punish ourselves:

- "I can't feel rage and hatred toward my mother. I owe her everything."
- "I can't ever feel overwhelmed by grief and despair. That would be pathetic and humiliating."
- "I can't feel sexual attraction toward that person. That would be totally perverse, inappropriate, and just wrong!"
- "I would never think something sexist or racist. What am I, a monster?"

When we feel the feelings we've built our defenses against, we typically respond with intense self-judgment, telling ourselves to "get over it" or that we're a horrible person. Yet *all* feelings are a normal and basic part of the human condition. Censoring and judging thoughts and feelings usually makes us feel worse — and it certainly doesn't make those thoughts and feelings go away. Far from it. Psychotherapists and meditation teachers understand that we are healthiest when we accept — and then allow ourselves to experience — all of our thoughts and feelings, without judgment and with compassion.

The best approach, in almost any situation — even the most painful one — is to recognize what we are feeling nonjudgmentally; to consider the situation, including how our values should inform us; and to then act based on a synthesis of our feelings and more rational evaluation. When we deny our feelings, we can't reach this synthesis, and we stay stuck in the feelings that we wanted to deny. This "stuckness" is the ironic reason that people who deny their feelings often end up being the *most* dominated by their emotions.

Psychotherapists and meditation teachers also know that you can't selectively filter out certain feelings without generalized consequences. If you are blocking anger and grief, for example, you will not be able to fully experience love and joy. In fact, it's not the content of our negative and painful thoughts themselves that matters — what matters is the way we face and process those thoughts and feelings and, ultimately, the way we act. You've probably already noticed that your negative and painful feelings exert power no matter how much you try to deny, punish, or tell yourself to "get over" them. But they do something else, too. When, for example, you tell yourself that you're a terrible person because you had a racist or sexist thought, you fail to explore, understand, and challenge your prejudiced ideas, inadvertently short-circuiting what could have been a self-reflective change process.

Although humans are masters at avoiding painful and uncomfortable feelings, we can learn to face and accept them by practicing self-compassion and by building "affect tolerance." This is what psychologists Kerry Kelly Novick and Jack Novick call "emotional muscle," and building it is a lifelong undertaking.[64] When you process your feelings in a healthy way — facing, accepting, and investigating their presence instead of attempting the fruitless work of repression — you allow and accept your whole self. Expanding your emotional range in this way offers at least two benefits that are key to living a fulfilling and moral life: It allows you to live in reality, without distortion, and it allows you to fully experience love, joy, and other "positive" emotions. Both of these benefits help you to better engage with emergencies — even emergencies at the scale of the climate crisis.

So, to begin building emotional muscle, you must first feel your feelings. This means you must try not to avoid negative or painful feelings. Instead, approach your pain with an attitude of curiosity and self-compassion. Although this isn't easy to do, almost everyone already has a powerful template for practicing such self-compassion. According to Kristin Neff, associate professor of Human Development and Culture at the University of

Texas at Austin, offering ourselves compassion means offering the same nonjudgmental comfort we offer our closest friends.[65] To build your emotional muscle, practice treating yourself and *all* of your feelings just the way you would treat a beloved friend who came to you for help. You wouldn't tell them to ignore their pain or call them a bad person. You would instead realize that they needed your help and — if you're a good friend — you would respond to them with compassion and empathy.

Our inability to practice self-compassion results in avoidance and intellectualization. This is why we so frequently classify anything related to the climate emergency as a "science" issue. By "science," we generally mean a boring, complicated issue that regular people can't understand. "Science" helps keep the climate in the intellectual realm, ensuring that we remain relatively safe from any difficult emotions. I know how it works because I used this defensive maneuver for years. I relegated climate to the "science" section of my mind, classifying it as "boring" and "for experts," and thus safeguarded myself from my fear associated with global warming. Instead, I resorted to the safer, more controllable emotion of resentment. I was annoyed at the idea that I was supposed to learn about something so seemingly arcane that it centered on a couple of degrees and some melting ice.

However, as I began to loosen my defenses, confront climate truth, and feel the pain associated with its acknowledgment, I began to open myself up to the reality that the climate emergency was not a "science" issue, but an *everything* and *everyone* issue. I began to see that the climate intensely and profoundly impacts all of humanity. And when I allowed myself to see the ways that the climate is a part of everything, I was able to recognize that it had been deeply affecting my heart all along.

Once I began to acknowledge the climate's comprehensive impact, I was able to free myself to fully feel the fear and pain that I had been repressing. It felt like the world was collapsing in on me. But it also felt deeply liberating. I was finally confronting the grief, apocalyptic fear, anger, and guilt that I had been working so hard to deny. Rather than relegating them to the corner of my consciousness, where they continued to nag at me, I put those feelings front and center, treating them — and myself — with compassion.

Classifying the climate emergency as a simple science issue may appear to free us from the messiness of our human feelings, but it's a trap. When the climate is confined to the intellectual realm, it only *seems* as though we absolve ourselves of discomfort; instead, we just repress our feelings and thereby limit the range of our response. Remember, any physical phenomena can be described solely in scientific terms. Look at cancer: A huge body of data and a scientific apparatus help to define cancer, but we consider a cancer diagnosis to be a highly personal, multidimensional emergency. When someone we love — or even someone we just know — is diagnosed with cancer, we don't say to each other, "Well, okay, this is an issue for science to solve," and turn away. Instead, we feel intense feelings of fear, pain, anticipatory loss, and grief, and we look for ways to help. We know that for the cancer sufferer, integrating the diagnosis will be an involved and challenging emotional process. We also know that when the cancer sufferer doesn't emotionally process their illness, it's a major problem that can impair treatment.[66]

Just as we can grasp the far-reaching impact of a cancer diagnosis without being oncologists or cancer researchers, we can grasp the impact of our climate emergency without being climate scientists. To respond fully and humanely to the climate emergency, we need only understand the basic concepts of the crisis and its implications, allow ourselves to face and feel the feelings we're avoiding, and then act.

In addition to practicing avoidance, denial, and repression to avoid feeling our feelings, we also use dissociation. Dissociation disturbs the normal links between thoughts, feelings, and actions, and can range from the mild "zoning out" to out-of-body experiences, amnesia, or, in extreme cases, even multiple personality disorder. Dissociation is a last-ditch defense that occurs when a mind is overwhelmed; it is frequently experienced during traumatic situations. Dissociation protects us during moments when nothing else does — for example, children often dissociate when exposed to violence. But dissociation isn't a healthy long-term strategy. When you dissociate, you lose touch with what is *actually* happening around you and *what you know is true.*

When we dissociate from the pain of the climate crisis, we protect ourselves from feeling the discomfort of pain, but we *do not protect ourselves from catastrophic climate breakdown.* Instead, when we dissociate to manage our feelings of overwhelm, we become stuck at the level of apathy and willful ignorance. We may protect our feelings, but we do so at the expense of our actual physical safety.

As you undertake the work of facing climate truth and begin to work toward transformation, you may feel like you're constantly fighting the desire to dissociate — to watch TV, to just zone out, to get drunk, or to self-medicate in some other way. It's true that facing the reality of danger and destruction on this planet is terribly painful. But allowing and accepting the pain is necessary for the work ahead. If you seek to move forward in reality, to help restore a safe climate and protect the world, you must allow yourself to experience the grief, terror, guilt, rage, and any other feeling that the truth evokes.

Facing climate truth and feeling the feelings associated with it require a personal transformation. The insights and guidance of psychoanalysis provide critical tools for this. Psychoanalysis is one of the great breakthroughs in the 20th century, on par with computers and antibiotics. I wish that everyone in the world could have access to high-quality psychoanalytic therapy. For me, psychoanalytic psychotherapy has helped in every area — and almost every stage — of my life. In my darkest times, therapy kept me afloat; in better times, as I sought to transform myself into a climate warrior, it supported me in reaching higher levels of insight, affect tolerance, acceptance, and self-compassion.

Please consider whether psychotherapy might be able to help you. You'll be in good company. Some outstanding social movement leaders have found help in psychotherapy. For example, AIDS activist Larry Kramer, discussed in Step Four, says that his years of psychoanalysis during college helped him confront and channel his own anger, fear, and grief into effectively responding to the AIDS crisis when others were overwhelmed.[67] Pope Francis, who has been a strong moral leader on the climate crisis, also underwent psychoanalytic psychotherapy in his 40s. According to Pope Francis, the therapy helped him "clarify a few things."[68]

Think of therapy as hiring a personal trainer — but one who helps prepare you for the marathon of life — and this is particularly true for your life as a climate warrior. I know that it's not always possible to consult or hire a professional. I am privileged to have been able to afford and have the time required for therapy. Readers who are prohibited by cost but not by time, and who are interested in pursuing psychotherapy might consider a lower-cost option at a psychoanalytic institute. Such institutes are staffed by advanced trainees with mental health degrees who are pursuing psychoanalytic training. Advanced trainees are often willing to take on clients at a reduced rate.

However, you can reap some of the benefits of psychoanalysis, even without a therapist's guidance and support. You can, for example, practice responding to your thoughts and feelings with curiosity rather than judgment. As described above, in this practice, you work to *stay with your* feelings. You *don't* downplay or ignore them, or attempt to intellectualize them so as to reassign their origins and effects. Instead, when you feel uncomfortable feelings, you assume an attitude of active interest and self-compassion.

To offer a sense of how this might work, let's try it out. Let's say that reading this book makes you feel overwhelmed, helpless, and cynical. Maybe you're so upset by my words that you're contemplating throwing this book out the window. Maybe you're starting to think some really negative, angry things about me.

Despite the naturalness of this kind of reaction, most readers have trouble responding to adverse thoughts with understanding. For many readers, self-judgment kicks in, harshly. They register their negative feelings toward this book or me and tell themselves that they're pathetic. They may berate themselves, saying something like, "You say you care about the environment. How can you give up like this? Are you weak? A coward?" Other readers work hard to shut out their negative feelings. Still other readers will argue with themselves, questioning their own intelligence or authority, or reproach themselves for their feelings.

But what happens when, rather than trying to prevent the negative feelings you may feel as you read this book, you assume an attitude of active interest and nonjudgmental curiosity? They are, after all, *just feelings*. They don't hurt anyone or have any impact in the world — only your actions do. Take a few deep breaths. As you become aware of negative feelings, such as a desire to deny or an urge to defend or a need to express skepticism, don't try to talk yourself out of these feelings. Instead of judging your feelings — and thus yourself — try to be curious about them, allowing yourself to really and fully experience them. Try, for example, to name the feeling — not justify, evaluate, or rationalize, just name: "This book is making me feel angry." Notice where in your body your anger is being experienced. Are you holding tension in your shoulders? Are you clenching your teeth?

By exploring your feelings, you become more comfortable with them and you build your emotional muscle, increasing your ability to face other, more painful feelings, such as those you will feel when you fully face the climate emergency and follow the call to protect humanity.

There are other approaches to learning self-compassion, increasing your affect tolerance, and building emotional muscle.

For example, you can identify the friends or family who are most comfortable talking about their feelings and who make you feel comfortable and accepted when you talk about yours. Make a regular effort to seek out these people to talk to about your feelings and theirs. Most of us, especially those of us not in therapy, are not in the practice of talking about the way events and people make us feel. We live in a culture that frequently rewards surface pleasantries on the one hand and smart judgments on the other.

But when you practice talking about your feelings, you get more comfortable noticing, identifying, and tolerating them. Start with relatively easy feelings: "I felt hurt and disappointed when my friend canceled the party," or "I feel worried and anxious about my upcoming test." Getting comfortable identifying and sharing these kinds of feelings, and having them received nonjudgmentally, prepares you to share other, more difficult feelings; it also prepares you for much more challenging work.

Mindfulness and meditation offer another approach. Clinical psychologist and Buddhist teacher Tara Brach discusses the benefits of meditation, using the acronym RAIN as a guide:[69]

- R: *Recognizing* what is happening.
- A: *Allowing* life to be just what it is.
- I: *Investigating* inner experience with gentle attention.
- N: *Nurturing* ourselves in our experiences.

RAIN helps its practitioners stop seeing themselves as being *the same* as their feelings. Our feelings do not define us; they happen inside of us, and we can nonjudgmentally notice them and choose how we react to them.

There are many other tools and practices that can prepare you to face your feelings and ready you to transform through climate truth. You might, for example, consider keeping a journal or log of your thoughts and feelings. For many people, this work can be another way to gain access to RAIN-based insights. By putting your feelings on paper, you create the distance that can help you accept your feelings, no matter what they are, with self-compassion.

I recommend you try to get comfortable with crying. This is often challenging, especially for those who have been taught that crying is a sign of weakness or that it signals an inability to cope. Crying is a specific act of emotional recognition and response; it is powerful, healthy, and necessary. It provides an outlet for all of the grief and pain inside you, helping link the emotional and physiological. When you cry, you release toxins and stress hormones from your body.[70, 71] Simply giving yourself permission to freely cry can be a tremendous relief[72] and can allow you to gain access to other repressed or ignored feelings.[73] Further, crying plays a critical social function, communicating to others, such as your friends and family, that you could use their comfort and support.[74] How will you know when you're making progress? You'll know because you'll allow yourself to experience more thoughts and feelings without judgment and censorship. You should feel *proud* when you notice yourself experiencing feelings that you are uncomfortable with — especially feelings that make you feel pathetic or guilty or otherwise vulnerable. If you have trouble crying, honor and praise yourself when you do cry. It is truly a victory. Every time you allow yourself to feel hard feelings, you expand your ability to tolerate affect. You build your emotional muscle. You make yourself stronger.

Feeling your feelings and practicing self-compassion allows you to access a greater range of feelings about the climate crisis. Living in climate truth means facing the truth of the climate and ecological emergencies emotionally, telling this truth publicly and privately, and allowing this truth to guide our actions.

Living in climate truth is hard, but it's the only way forward. You *will* experience a wide range of emotions, and they *will* be intense. They may be irrational, ugly, or upsetting, but they may also be wholly appropriate to the stakes of a crisis that is set to lead to mass destruction and death. Your painful feelings spring from the best parts of yourself, from your empathy, sense of responsibility, love for others, and love of life. These feelings connect you to all life and will fuel the work ahead. Immersing yourself fully in them is a heroic, even sacred, undertaking.

The work you do in learning to accept the feelings that arise in your personal life will enable you to accept, to feel, and to use the intense emotional reactions that will result from living your life in climate truth. When you accept and process your painful feelings in your personal life, you are also better able to access, accept, and process the pain of others. The more comfortable and confident you are with fear and pain, the more you will be able to help others accept their own intense feelings and turn them into action. This, too, will be necessary because, as we must strive to remember, we are all on this rollercoaster together.

Given the realities of our situation, we have a duty to grow as quickly and effectively as possible as individuals to most effectively acknowledge climate truth and position ourselves to respond — with others — to the crisis. Any meaningful, world-transforming social movement is comprised of hundreds and thousands of individuals who have changed their outlook and their behavior and are leading the way in creating a better world.

Even though I have been in therapy for more than a decade, and I am generally quite emotionally open, I have had to work long and hard to accept all the painful and difficult feelings the climate crisis has stirred up in me. My fear is a constant presence in my chest and stomach. I am afraid for myself and my family, and even more for the world as a whole. I am afraid of the unbearable pain I will feel as civilization continues to collapse; as communities, institutions, and states fail; as more and more people are immiserated, as species become extinct. But fear has become my greatest motivator, urging me on.

Being motivated by fear can help me keep other motives, such as my desire for narcissistic gratification, in check. I am a competitive person, so concerns about "getting credit" or being the "best" or "directing the most popular" organization nag me. However, when I feel envy or competitiveness toward another organization or climate warrior, I don't deny my feelings. Instead, I acknowledge them and then ask myself an honest question: "Do I want to be a big shot, or do I want to prevent collapse? Which is my most important priority?" My answer is always the same, and I am made to remember how small and irrational my narcissistic ego needs are. Note that I do not judge myself harshly for feeling competitive or ego-driven. It's fine. It's human. I simply remind myself that those feelings are not in line with my values and priorities, and I refuse to be driven by them. If I instead denied my competitive feelings or judged myself for experiencing them, I would be much more likely to act from feelings of competition and judgment.

In addition to fear, I feel a seemingly bottomless sadness. I am heartsick about the ecological crisis. There is so much suffering in the world *now*, and we are heading straight into total devastation. Human life and our natural world are the greatest blessings imaginable. This is true whether we believe life was given by God, by other spiritual forces, or by randomness. The most intelligent species is destroying itself and bringing on a sixth mass extinction. We are destroying our greatest gifts. We are choosing death.

I cannot be cynical in the face of all this loss and suffering, telling myself that humanity is irredeemable and collapse is inevitable. For me, this tragedy is all the more painful because I believe in the immensity of untapped human potential. I know that growth and change are possible, both in individuals and in societies. In fact, as a psychologist, I believe that the tools that I've been talking about — processing and accepting feelings with nonjudgmental self-compassion — are key to achieving growth and transformation. Humans are particularly excellent at responding collaboratively and effectively when they face existential crises. That is why the gap between who we are and what we're doing now and what we *could* be and do is so devastating. We are capable of so much more than this.

I also feel disgust, shame, and contempt. I am disgusted by our materialistic, racist, alienating, and dehumanizing economic and political system. I am disgusted and ashamed of myself for taking part in it. I feel disgust and contempt for everyone else who takes part in it. Every day, we pump more carbon into the atmosphere and put more plastic into the ocean, and for the most part, we do it with a smile. I sometimes wonder if, when we arrive at the pearly gates, and St. Peter is determining whether to let us into heaven, we will be faced with a pile of garbage and a CO_2 calculation: Time to tally up all the plastic crap and everything else we ever threw "away." Even worse, we are squandering precious time to transform, which is an unforgivable crime. It makes me nauseous.

I feel anger and rage. I feel angry and exasperated when I see people idling their cars or posting about their trips to tropical places. I feel angry when people just live their lives and don't fight the climate crisis. I am angry at the Democratic Party and the progressive movement for treating the climate crisis as a side issue for so many years. I feel angry that the environmental movement hasn't had more success, and I am full of critiques about their focus and goals. I am angry at Baby Boomers — my parents' generation — for allowing our crisis to reach this point.

Although I know some people feel visceral anger at oil company executives or GOP politicians, I've always accepted that evil people do evil things. I agree they have committed crimes against humanity and should be tried at the International Court of Justice. However, I just don't feel angry at them. Rather I feel angry with people I know — and often people I love — for failing to protect me and all life. I feel betrayed by my family members who voted for Trump. But I am also angered and let down by those who support my climate activism as "my thing," but don't recognize that it needs to be "their thing," too.

And sometimes I feel what I consider destructive glee. I feel so angry about collective denial — and so disgusted by our death-machine culture and economy — that I think something like, "Don't come crying to me when the crisis is at your door. You had your chance; I warned you about this." It's kind of a ghoulish self-righteous revenge fantasy. In this feeling, being "right" becomes more important than being safe.

I also feel guilt. I feel guilty for not doing more to solve the climate crisis: not working harder, not reaching more people, not donating more money or otherwise sacrificing more. I feel guilty about all of the horrible things happening in the world, all of the present suffering and oppression that I am not focused on because I am focused on preventing a catastrophic breakdown in the near future. I feel guilty for my privileges and life of comfort, and for my consumption. When I drive, eat meat or dairy, use plastics, or take a Lyft, I feel guilty, knowing I am contributing to the destruction of the world. These feelings make me imagine living like Gandhi or a monk, and I want to renounce everything in this fallen world.

I feel morally responsible for doing everything I can to prevent the full unfolding of the climate crisis. I wake up feeling the heaviness of this responsibility. It stays with me all day. Although I can sometimes relax and socialize, I always feel that those pursuits are secondary — and in support of — my true mission. Although I feel tremendous pride in TCM's accomplishments, until the American economy is fully mobilized and we are moving rapidly toward global mobilization and restoring a safe climate, I do not feel that I've met my responsibility.

I often feel despair and helplessness. Sometimes I am without hope. I feel that my efforts, and the efforts of many other passionate and dedicated individuals, are doomed to fail. It is too late; humanity is irredeemable. There is nothing I or anyone else can do.

Sometimes, I want to die. My despair, terror, and guilt about the collapse of civilization and mass extinction of species have made death seem like sweet relief. For some readers, this admission might sound shocking, but most people I know who are living in climate truth grapple with this feeling at times. Frankly, it's an understandable response to the crisis. We are living in a broken world, in an age of mass violence and death. The crisis is global, and there is no safe haven. Death can sometimes feel like the only way to end the nearly unbearable feelings and escape the hellish future we are careening toward. In this instance, it is *especially critical* to distinguish the difference between thoughts and feelings, and actions.

As with all feelings, it is healthier to nonjudgmentally acknowledge the feeling of wanting to die than to attempt to deny it. However, no one should kill themselves in response to the climate emergency. Not while there is still any chance of restoring a safe climate. We should choose the more challenging path of going "all in" to solve the climate crisis. Let the pain be your fuel. Let your total rejection of the status quo give you the courage to transform your life, to stand out from the crowd, and demand transformative action.

And if you do notice any thoughts of suicide or self-harm arising in regards to the climate crisis (or anything else), please don't hesitate to contact a mental health professional or call the National Suicide Prevention Hotline at 1-800-273-TALK if you are in America. Revealing these feelings to someone else can be terrifying, but it will help. Going "all in" does not mean going it alone.

I feel alienation. I feel so different from people who aren't living in climate truth. When I walk down the street in Brooklyn, and everyone is going blithely about their lives, I feel very strange and very different from everyone. I feel they cannot really understand me and that I cannot really understand them — and what I see as their petty or self-involved concerns. On the other hand, I cherish relationships with people who are living in climate truth.

For me and others who are dedicated to TCM, it feels like an amazing and huge relief to see, connect, and work with others who live in climate truth.

Sometimes, especially when surrounded by allies and when our work is bearing fruit, I am bursting with hope for a new world that is coming. I feel that humanity can and will respond to this crisis. I believe that once we are in emergency mode, we can accomplish shockingly ambitious feats. I believe we will transform into a species that treasures and nurtures all life. When I look at history, I see extraordinary social movements that have sparked dramatic political shifts in consciousness and morality. I see what American industry and society accomplished during WWII. Then and now, we can wake up from the trance of denial and go all in to fight an existential threat. I feel proud to be a part of it.

Of the range of emotions that I describe, two deserve special consideration: fear and grief. While *all* feelings should be non-judgmentally welcomed with self-compassion, fear and grief are especially potent in transforming yourself and society, and avoiding them has been especially destructive to the climate movement's efficacy.

Fear is one of the seven primal emotions that mammals experience, and one of the most reliably effective motivators for the entire animal kingdom. It evolved in animals to force a response to threats.[75] Fear helps us protect ourselves; it mediates between perceiving danger and taking defensive action. It is literally the motivator that commands our response — without it, we can't react.[76]

That's why, among the many and varied emotions we are suppressing about the climate crisis, fear is the emotion we need to feel the most. Fear is the mechanism through which we turn the perception of danger into self-protective action — without it, we expose ourselves to terrible risks. Telling the whole, frightening truth is the climate movement's most underutilized asset, and it will unlock tremendous potential for transformation *if* we provide a heroic solution that we are all invited to take part in creating.

Feeling your fear and grief is going to hurt, but don't worry, the pain is there for a reason. Also, it's not new pain. It has been with you your whole life — no one living on this planet can avoid the ecological crisis entirely. Practicing the self-compassion that will allow you to feel, and keep feeling, the fear and its resulting pain will propel you forward into committed action; it will help you go all-in as a climate warrior.

When I began to learn more about the climate emergency, I felt despair. I knew that unprocessed despair can be paralyzing in its power. I knew I needed to grieve and mourn and work through these feelings. From a psychological perspective, I also knew that the grieving process is the only healthy response to incalculable loss. It's seldom discussed outside a therapist's office, but grief is the best method we have for reacting to meaningful losses and adapting to new realities.

Therefore, once we face the truth and feel our feelings associated with the climate emergency, we must allow ourselves to experience and work through our grief. Grief is painful, and for some readers it might be the most painful feeling yet. But if we stop ourselves from feeling grief, we stop ourselves from emotionally processing the reality of our loss. As we already know, if we can't process reality, then we can't live in reality: We become imprisoned and immobile. Grief ensures we don't get stuck in denial, living in the past or in fantasy versions of the present and future. Think of the widower who cannot acknowledge the death of his wife, who cannot cry, and who never cleans out her closet. He remains stuck, suspended between the past and his fantasy version of the present. Since he can't grieve, he can't adapt to his new reality and cannot find satisfaction at all in what is a new — if unwelcome — stage of life.

According to Joanna Macy, an environmental advocate and spiritual teacher, we must grieve to accept the reality and pain of loss. In her book, *Active Hope: How to Face the Mess We Are in*

Without Going Crazy, Macy writes that grief follows from a heart that has been broken by loss. The initial task of our grief is simply to help us accept the actuality of the loss. The second task of our grief is to — again — allow ourselves to feel pain, which helps us understand the meaning and depth of our loss:

> When we feel this emotion [of grief], we know not only that the loss is real but also that it matters to us. That's the digestion phase — where the awareness sinks to a deeper place within us so that we take in what it means.[77]

Like McKibben, Macy argues that we must acknowledge the truth of reality before we can move on. But Macy argues that we acknowledge reality through grief. She says that it is grief that helps us to "find a way forward that is based on an accurate perception of reality."[78]

When we avoid the hard work of grieving and mourning, we also avoid the opportunity to feel not just what we've lost *but what we love*. When I allowed myself to grieve the losses we've already sustained in the climate crisis, I was able to recognize the climate emergency as a substantive part of my reality, a major aspect of my life. Feeling grief reinforced my recognition of the seriousness of the climate emergency and the seriousness of our collective loss. But grief did something else, too. It reminded me of how much I love this world. The depth of my grief was a direct response to how connected I am — and want to be — to the living world. It gave me a new breath of life.

This makes grief not only worth experiencing but worth honoring. If you're reading this, you likely already know that we've lost much. Millions of people have already died because of the climate and ecological crisis, mainly because of hunger and communicable diseases.[79] Most are among the world's poorest people. We grieve them because they matter.

It's not just humanity that has been lost. Biodiversity — the riot of life — is also being speedily destroyed. We have already seen thousands of species slip into extinction. The species that still exist are rapidly losing numbers. Nisha Gaind, reporting in the World Wildlife Fund's *Living Planet Report 2016*, writes that the population of vertebrates has declined 59 percent since 1970.[80] These losses are accelerating.

Although it is necessary to grieve, grief is seldom experienced in any straightforward way. Author, editor, and climate-justice essayist Mary Annaïse Heglar captures its nuance when she writes:

> I floated around on a dark, dark cloud. I frequently and randomly burst into tears, and I'd refuse to admit to myself that I knew exactly why I was crying. When I was around bustling crowds of people, I saw death and destruction. When I walked on dry land, I saw floods. I imagined wild animals, especially snakes, getting out of the zoos in the aftermath of natural disasters. I worried about how we would treat each other in the face of such calamity.[81]

Do you feel that pain? I think that you do, though it may be too diffuse and unnamed to notice. We must decide that these losses deserve to be remembered, felt, and mourned. We must recognize that only by grieving these deaths and extinctions can we fully process our pain, honor our loss, and enable ourselves to engage in the reality of our already-diminished collective.

I must point out here that grieving is not the same as giving up. There is a strain of climate "doomers" who say that humanity and the natural world is a lost cause. They believe that our Earth is in hospice and we should prepare it, and ourselves, for death.[82, 83] These doomers see their current calling as simply expressing their grief and persuading climate warriors that they are high on "hopium." While doomers understand the need for grief, they misunderstand grief's purpose. They use grief as an endpoint, an excuse for inaction. Instead, grief reinforces our ties to humanity and to life, and calls us to *fight* to save as much life as possible and — hopefully — to restore what we have lost.

When you acknowledge the truth that we are facing a climate emergency and allow yourself to feel the all-encompassing grief of what we've already lost, you also begin the important work of re-establishing your visceral and intimate connection to all life. Despite the fact that we've been taught to view the natural world as fundamentally separate from humanity — a "resource" for our use — and to ignore our fundamental connection to the other living beings of the planet, the connection is our natural, spiritual birthright. Macy cites Zen Buddhism to argue for the necessity of the connection:

> The Vietnamese Zen master Thich Nhat Hanh was once asked what we needed to do to save our world. "What we most need to do," he replied, "is to hear within us the sounds of the Earth crying."[84]

But, writes Macy, the Earth can cry within or through us only when we stop viewing ourselves as separate individuals and start viewing ourselves as part of a collective:

> If we think of ourselves as deeply embedded in a larger web of life, as Gaia theory, Buddhism, and many other, especially indigenous, spiritual traditions suggest, then the idea of the world feeling through us seems entirely natural.[85]

This is a very different view of the self than what Macy calls the "extreme individualism" that "takes each of us as a separate bundle of self-interest, with motivations and emotions that only make sense within the confines of our own stories."[86] When we throw out the lessons of neoliberalism and instead allow ourselves to face the truth of the climate emergency and grieve its associated losses, we begin to hear the Earth's story, which is a story of our interconnectedness.

Acknowledging our connection to the Earth also means grieving *our own futures* — the futures we had planned and hoped for, the ones we wanted to believe we could count on. We must recognize that our lives will no longer unfold on a stable planet. We must see that we are not separate from the natural world but that our lives — today and tomorrow — are totally dependent on a steady climate and civilization: If the oceans die, we die. If the forests die, we die.

Although the loss of our future hopes and plans may feel abstract and hard to access, this contemplation is a necessary part of the grieving process. Heglar speaks to this grief on her road to becoming a climate warrior: "I'd silently been asking myself: What am I fighting for? What am I trying for? Why am I paying my student loans? Hell, why am I saving for retirement?"[87]

I experienced something similar. When I was a child, my mother told me that I could be "anything I wanted to be." I knew this wasn't literally true, but I also knew that I had many options. As I grew up, I set out to become a clinical psychologist, with plans to write books about psychology for popular audiences. I imagined myself married with children. What a lovely life I had planned! It was going to be meaningful, intellectually stimulating, financially rewarding, and rich in relationships.

But when I forced myself to face the climate crisis and to accept the truth of the climate emergency, and when I started the process of grieving what was already lost, I realized that my lovely life was based on a lie. It was not going to happen. Maybe I could still pull off living my perfect life — at least for a decade or so — but would it be satisfying if it was unfolding while tens of millions of refugees streamed out of regions made unlivable by heat, drought, or flood; state after state failed; and collapse seemed imminent?

Ultimately, I had to acknowledge that the future I had planned was ruined. I was never going to be able to lead a happy and satisfying life while watching the world burn — I was already too connected to other people for that. I knew I had to say goodbye to the future I had planned for myself. In many ways, I had to say goodbye to the person who had made those plans, and so I grieved those losses, too.

In addition to all the other ways that the climate emergency has impacted us, it threatens our sense of the future because it forces us to question our assumptions about progress — about our belief that "the arc of the moral universe is long, but it bends toward justice."[88] The climate emergency has the power to set back thousands of years of human development. Indeed, it has already challenged our basic sense of linear time and the future's unfolding. Wallace-Wells describes this phenomenon in his book, *The Uninhabitable Earth*:

> Early naturalists talked often about "deep time" — the perception they had, contemplating the grandeur of this valley or that rock basin, of the profound slowness of nature. But the perspective changes when history accelerates. What lies in store for us is more like what aboriginal Australians, talking with Victorian anthropologists, called "dreamtime," or "everywhen": the semi-mythical experience of encountering, in the present moment, an out-of-time past, when ancestors, heroes, and demigods crowded an epic stage. You can find it already by watching footage of an iceberg collapsing into the sea — a feeling of history happening all at once.[89]

I feel this collapse of time every day: My hopes, dreams, and plans for the future are irrelevant, and the *now* has become almost unbearably important. This is not hyperbole. My actions, like yours — this year and next — will have an incalculable impact on all life.

While grieving the loss of the future I had planned was painful, it gave me something important, too. I gained the ability to engage more fully and meaningfully in the here and now of reality and morality. I also gained the freedom to change course and envision a new future. I was able to set out on a new path, with a future dictated by the personal mission to do whatever I could to initiate a global response to the climate emergency. Letting go of my hopes and plans, which in themselves constituted a kind of climate denial, allowed me to live in line with my values, which meant living in climate truth. This gave me a powerful hope — one that I had never felt before — and has motivated my transformation into a climate warrior who is prepared to do everything I can to prevent catastrophic outcomes from fully unfolding, and help restore the health of the climate and protect all life.

You, too, must grieve. Only when you are able to face the future as deeply imperiled — not the reliable, stable future you were promised and imagined — will you be ready to move on. You must let go of denial and face climate truth. You must acknowledge the present reality of our climate emergency. You must grieve the people and biodiversity we've already lost and the lives we lose every day. You must grieve the future you've dreamed of. This will allow you to step into the now and motivate you to act. In this way, you can turn grief into power, rather than stay stuck in paralytic despair.

You are probably hurting after reading all that. I know this is hard, and that's why I want to say thank you for having the courage to face the truth with your heart and mind. It matters a great deal to our collective chances. If it is any comfort, feeling your feelings, facing the truth, and grieving our losses are the hardest parts of your transformation. It should get a bit less painful and a little more uplifting from here. Onward!

Questions for Reflection and Discussion

- Do you judge yourself harshly for what you think and feel?
- Which emotions are the most challenging for you to feel? Do you criticize yourself for having them and try to shut them down? (For example, "Oh, don't feel angry and envious; you are such a bitch!" or "Boys don't cry.")
- Have you felt viscerally afraid of the climate emergency? What visual image or idea triggers your fear?
- Have you ever been in a life-threatening emergency? Do you remember the mental experience of your total focus on achieving safety?
- What are your cherished hopes and plans for the future? Have you integrated your understanding of the climate crisis into those plans?
- Have you experienced grief for the people, species, and ecosystems already lost? If so, describe those feelings.
- Does your experience of grief motivate you to protect life that is not already lost?

STEP THREE:
Reimagine Your Life Story

It is likely clear to you that to protect humanity and the living world you need to reconsider your fundamentals: Who are you? Why are you here? What is your purpose? The climate emergency challenges the answers we thought we had and invites us to revisit them anew. I have discussed the need to grieve the future you thought you had. But what can replace it? In the context of the ecological emergency, we must each revise our "story of self." In this new story, you are the hero. This designation might feel over the top. It might make you uncomfortable. But it's true: Humanity needs as many heroes as it can get. Will you join the team?

Everything in your life, including its most painful challenges, has prepared you for this role. Perhaps it has all led you here. You've experienced difficulties, but they've made you stronger. You've received gifts — whether a nurturing home, a strong moral sense, a good education, financial security, or specific talents — but they were not randomly given. Everything you've experienced and gained has prepared you for your mission.

I know it's true for you because it's true for me, and it's true for so many others who have faced the truth and reoriented their lives around the climate emergency. Alexandria Ocasio-Cortez was a bartender who, while protesting at Standing Rock, decided to run for Congress; she has succeeded in forwarding the climate emergency movement and drastically raising expectations of what is politically possible. Genevieve Guenther, a Renaissance literature professor, has become an incredibly effective advocate for persuading journalists to include the climate emergency in their stories. Trevor Neilson, businessman and co-founder of i(x) investments, founded the Climate Emergency Fund, which gives grants to the young organizations of the climate emergency movement, such as local XR chapters, school strikers, and Climate Mobilization Project. These individuals are taking their skills, talents, networks, and resources and re-orienting them towards solving the climate emergency.

I, too, have been given many gifts and privileges that I strive to put into the service of the mission, including access to a world-class education; years of intensive psychotherapy; the socioeconomic privilege that has allowed me to work for years as a volunteer; and family and friends who believe in and support me.

But I can see that the challenges I faced also taught me lessons to help me meet the challenge of saving the world. My childhood prepared me to take responsibility for difficult problems, to work through hard emotions, and to shed preconceived ideas. My parents both came from troubled homes — my mother was the daughter of teen parents, and my father was the son of a depressed, traumatized Holocaust survivor. My parents valued the empathy and emotional insight I expressed as a youth and encouraged me to take on a leadership role in the emotional and social realms in my family. For example, I remember them asking for input on how to raise my brothers and me, and taking my advice seriously. Perhaps this is why I have always been dubious of expecting authorities or anyone else to solve problems for me. It has always felt natural for me to take on a leadership position, and I have always been encouraged to say, "OK, I'll handle this."

I've also drawn conviction from my family's background. I grew up hearing about the Holocaust from my grandmother. When she visited my family, which was often, she talked almost exclusively about the Holocaust. She saw everything through its life-altering lens. She carried with her a deep and abiding feeling of betrayal — not just the betrayal she experienced by the Nazis but the betrayal she experienced by ordinary Germans like her schoolteacher, who refused to acknowledge her on the street after my grandmother was kicked out of school. Hearing these stories at every visit instilled in me a visceral understanding of the adage that "all it takes for evil to triumph is for good [people] to do nothing."

I've also faced personal tragedy. My first love was my high-school boyfriend, a brilliant actor, writer, and scholar who taught me how to play chess and, who, as my partner in our school's mock trial team, led us to win the state championship. We had been together for more than two years, thinking about our bright futures and considering attending college together, when disaster struck. He became floridly psychotic and experienced his first of many hospitalizations.

It is impossible to convey the overwhelming pain of watching a loved one become psychotic. In just a few days, my boyfriend went from being the most loved and trusted person in my world to being someone scary and confusing, someone who didn't always recognize me. He was treated and medicated, but he wasn't the same, and never would be again.

Neither would I. The pain I experienced watching him struggle in the grip of a disease he neither understood nor was able to fight effectively was a hurt I had never known before. College was supposed to be an exciting and emotionally connected experience. I was supposed to be spreading my wings. Instead, I was alienated, traumatized, and depressed. During my first year of college, my boyfriend was in and out of the hospital; I felt tremendous guilt about leaving him in such a terrible condition. In some ways, I felt like the proverbial wolf that had been caught in a trap: I felt like I was chewing off one of my own legs to get away and survive. I wondered if I would feel whole, or good about myself, ever again.

Of course, my boyfriend's life was immeasurably harder. His illness never relented, and each psychotic episode devastated him, to say nothing of the devastation experienced by his family and his friends. After struggling for over a decade with his disease, he reached a point of no return and killed himself. His death left a gaping hole in the lives of everybody who knew him, everybody who watched and worried as he struggled with a powerful disease. He was a unique and beautiful person, overflowing with love and talent. He is irreplaceable, and his illness and death were unfair. I miss him terribly.

The experience taught me so many things, but among the most important was the most simple: Terrible things — things you never thought to fear — actually happen. People may understand that bad things happen, but they think that bad things usually happen to *other* people. They may understand that things go wrong, but they think that when they go wrong, they go wrong in mostly expected ways. Only when someone has experienced a shattering tragedy or trauma do they realize that unforeseen, life-altering disasters can and do happen. I know in my bones that terrible things happen. My grandmother told me this over and over again, but I have also experienced it myself. Things happen that are so catastrophic you never even knew you should fear them.

My experience with my high-school boyfriend's tragedy taught me another lesson: what it feels like to possess a painful and uncomfortable truth. After his death, I received immense comfort from many people who reached out to me to tell me how sorry they were and that they were thinking of me. During his psychosis, however, I had received little of this support — people didn't want to talk about what was happening to him. They didn't know what to say; they feared saying the wrong thing; and they wanted to put their hope into a future where he was fine. The result was that I felt very alone.

Isolation is a common experience for people whose romantic partners experience a psychotic episode. More than five years after I watched my boyfriend turn into someone entirely different, I was still searching for answers about what had happened — to him and to me. In my effort to understand my trauma, I developed my doctoral dissertation on the traumatic impact of psychotic episodes on romantic partners by interviewing women who had witnessed their partners' psychosis. Two key themes emerged from my work: the tendency of friends and family members to downplay the severity of an illness, and the estrangement and alienation the women felt because, in their words, "no one understood" and there was "no one to talk to."[90]

As we've seen with the denial surrounding our climate crisis, when an experience or phenomenon is taboo and shrouded in silence or shame, it becomes even more powerful and destructive. This is why, in Alcoholics Anonymous, they say that "you are only as sick as your secrets." In my interviews, many of the women were relieved and unburdened to be able to talk with someone about their experiences.

Brené Brown, a well-known author and professor at the University of Houston, writes about the horrible consequences of shame. She encourages readers to have the courage to be vulnerable and share their lives with trusted others, especially the parts of life that make them feel ashamed. When met with empathy and understanding, our shame dissipates; connection and wholeness take its place. When I was a therapist, I worked with my clients to discuss the most painful parts of their lives — things that had made them feel terribly ashamed — and I saw the healing power of taking painful secrets out of the shadows.

I do this same work with the people who join monthly phone calls with TCM, discussing how it feels to live with the truth of the climate and ecological crisis. "No one understands" and "I feel so alone" are prominent experiences for those who are grappling with and working to face the climate crisis. Helping someone face and work through their personal crises and helping someone face and process the climate crisis aren't very different. In both contexts, the helper must express empathy, curiosity, and a willingness to tolerate painful feelings in service of the truth.

It has been redemptive to think of my experience witnessing my high-school boyfriend's psychosis as preparation for the mission of protecting humanity and all life. For years, I felt bitterness and resentment. I felt that the world had been terribly unfair to me. I wished I could just have a "normal and happy" life. But I don't wish that anymore. I've come to realize that because I experienced catastrophic collapse up close, I can see a collapse that others might be blind to. My experience made me more comfortable and motivated to break the silence and speak uncomfortable truths. While I am forever sorry that I couldn't protect or heal my boyfriend, I am glad to have an opportunity to prevent another calamity and help heal our society.

I am also grateful that my grandmother taught me about catastrophe at a young age, including giving me a moral framework around it. She taught me that in the face of imminent disaster, it is not morally acceptable to be passive. We cannot be the "good Germans" who, in their passivity and inaction, supported the Third Reich's atrocities. We cannot make "fitting in" or "not rocking the boat" our priorities. Rather, we must uphold our values — social norms be damned — and actively work to prevent catastrophe.

Although my story is unique, it shares the pattern of ups and downs that is common in most peoples' lives. Take a moment to reflect on your own life: How have your gifts and difficult challenges brought you to this moment, to this book, and to this cause? The reflection may reveal that you are more ready for this mission than you imagine. In fact, maybe you have been preparing for it all along. It's easier to see that the emotional and practical skills that you've developed, the talents you've cultivated, and the relationships you've built will assist you on your mission. But it's your hardest and most painful moments that have prepared you in critical ways. Perhaps a childhood illness taught you about the vulnerability of all things and your capacity for healing. Maybe being bullied taught you to be assertive when necessary. Perhaps you experienced racism that taught you that the world is in desperate need of transformative change and that confidence has to come from within. Maybe the premature death of a loved one made you fiercer in your desire to protect others.

There is something else on your side, too. While I've considered myself an atheist for most of my life and I've never found solace in the belief in an all-seeing, protective God, my climate work has given me a profound spiritual sense. I'm not necessarily saying that God or Fate has guided my path, preparing me for this work, but I do believe there is an unseen, powerful force on our planet that wants to live. It's a force that exists inside all of us: It's the reason our hearts beat without our volition. It's the reason we're able to breathe without conscious effort. This vital force ensures that we, along with plants, other animals, and even microbes, cannot help but seek out hospitable environments in which to grow and reproduce. It is a product of evolution — plants and animals that don't display a drive to eat, drink, and avoid predators, die.

But this force also manifests in our *collective* desire to protect not just our own lives but to protect *all* life. Different cultures have recognized this essential component of human experience and expressed it in different ways — "Tao" in Chinese, "Tathāgatagarbha" (or Buddha-nature) in Sanskrit, "Ruh al-Qudus" in Arabic, "Chai" in Hebrew, "Holy Spirit" to Lutherans, "The Power of Love" to Martin Luther King, Jr., and "Spirit" to the Water Protectors at Standing Rock.

In his book, *The Social Conquest of Earth,* biologist E.O. Wilson describes something similar, writing that humans evolved both to maximize their individual self-interests *and also to protect the group and make sure it thrives.* These two forces — individual selection (our desires to protect and enhance ourselves) and group selection (our desire to protect and enhance the group) — have been key to shaping human evolution:

> The dilemma of good and evil was created by multilevel selection, in which individual selection and

> group selection act together on the same individual but largely in opposition to each other. Individual selection is the result of competition for survival and reproduction among members of the same group. It shapes instincts in each member that are fundamentally selfish with reference to other members. In contrast, group selection consists of competition between societies, through both direct conflict and differential competence in exploiting the environment. Group selection shapes instincts that tend to make individuals altruistic toward one another (but not toward members of other groups). Individual selection is responsible for much of what we call sin, while group selection is responsible for the greater part of virtue. Together they have created the conflict between the poorer and the better angels of our nature.[91]

Wilson points out that altruism forged by group selection is extended only to the "in" group. But that won't work this time. This time, we need to all act together for the good of all. But I believe that the force on Earth that wants to live is pushing us beyond that boundary, toward a realization of interconnectedness and the individual belief that each person's safety and well-being are intertwined with the biosphere's safety and well-being. As Martin Luther King, Jr., said in "Letter from a Birmingham Jail": "We are caught in an inescapable network of mutuality, tied in a single garment of destiny." [92] We will only be able to truly thrive on a healthy planet when we take seriously our responsibility to protect and nurture each other and the natural world.

This vital force is the liveliness we feel in ourselves, and our sense of connection to and love for other living things. It is the part of us that feels responsible for others and for all life. It is the part of us that makes existence worthwhile for its own sake. It is the part of us that makes us more than mere self-interested machines, programmed to consume, compete, and build wealth and power. It's the part that makes us unique creatures capable of protecting and contributing to the greater community of life.

As Wilson points out, there are other forces within us, too. The drives for greed, separation, and narcissistic gratification are also inside us. Buddhism regards greed and hatred as mental afflictions that veil and obscure our true life-affirming nature. Our consumerist culture and alienating political economy have nurtured these negative forces and quieted the force of Earth that wants to live, regarding it as naïve and undisciplined. The consumptive, narcissistic, necrophilic drives have been the dominant ones in American culture for decades. But — and here's the good news — the climate emergency movement suggests that the better angels of our nature, the ones presiding over the interconnected, life-loving, protector forces, are resurgent. I believe that the force that wants to live is making a last stand as the climate crisis approaches the point of no return, likely because it faces the same extinction as the rest of us: If we die, it dies. But it is not going softly into the good night. Instead, it has risen up in many of us, commanding us to drop everything and fight for life. Even better, once you listen to this force and decide to fight for humanity and all life, it can become an active — if silent — partner.

Once I decided to do everything I can to protect humanity and the natural world, the wind has been at my back. I have been helped by hidden hands: The right people have appeared at the right time, and they have helped push me forward. I didn't suddenly feel the life force all at once, however. When I started to read more about the climate emergency, especially *Eaarth*, the climate dominated my thoughts. It felt as though sirens were blaring inside of me. I had nightmares about huge waves. Mostly, I felt raw terror, paired with grief and sorrow. I had trouble focusing on the issues that had previously occupied my mind, such as my academic and career success, my patients, my social life, my appearance, or my apartment. It came to a point where I could think about only one thing: the impending disaster.

I knew I needed to do something. I started to write and publish, but it didn't feel like enough. It wasn't until I felt the life force in me that I began to hear more substantive answers. I clearly remember the moment I began to hear them most explicitly. On a winter evening in 2013, my friend Ryan Peterson asked me to set my sights higher than I would have ever set them myself: "Discourse isn't enough," he said. "Think — what could you do or what could we do together that could actually *solve* the climate crisis?" His words resonated in me, echoing back at a visceral level, and I heard then that not only was I connected to all of life, but that because of that connection I could — I had to — take responsibility for protecting the entire web of connection.

I had never let myself think so big; I had imagined I could offer commentary, not that I could actually try to *solve* the crisis. Yet, once Ryan had issued the challenge, I knew that I had never wanted anything more. The force on Earth that wants to live was coursing through my veins. Whereas the climate emergency once filled me with terror and despair, the challenge to "solve the crisis" lit a fire in me that has never dimmed. I have established a long-term partnership with the force on Earth that wants to live; I have joined its team and become one of its agents.

Taking responsibility for solving the climate emergency and founding TCM has changed many aspects of myself, including my identity, my priorities, what makes me feel good about myself, and how I spend my time. It is probably no surprise to hear that it has changed me, but it's also changed others who have joined our efforts. TCM asks our volunteers to go "all in" — not just for themselves but for everyone. By fighting for humanity and all life, we channel the forces within us of *both* individual and group selection. We channel the force on Earth that wants to live. We have realized that initiating a WWII-scale climate mobilization is the *only* way forward, and we are transformed by the responsibility of our mission.

Questions for Reflection and Discussion

- How has your life prepared you to face the climate emergency?
    - What challenges have you faced that have prepared you to tell the truth, be resilient in a catastrophe, or lead or help others?
    - What have you learned from your parents, grandparents, and other ancestors that has prepared you for the climate emergency?
    - Have you ever experienced or witnessed a catastrophic collapse? Did someone you know suffer a psychological breakdown? Did a neighborhood burn down, an organization destroy itself, a society descend into chaos? What have you learned from those experiences?
- Can you envision a life that revolves around a commitment to protect all life?
- What are your greatest fears about becoming a climate hero? What are your hopes?
- In what ways do you feel unprepared? What do you need to work on? What do you need help with?
- Do you have a spiritual practice and perspective that equips you to protect humanity and all life?
- Does the concept of a "force on Earth that wants to live" resonate with you? Do you feel that force inside of you?

STEP FOUR:
Understand and Enter Emergency Mode

IF THE PRECEDING MATERIAL unsettles and terrifies you, good. We are in terrible danger — you and I; my family and your family. Everyone and everything we love — all life — is in danger. Feel the fear, grieve the future, and let it motivate you to action.

If you have not experienced that fear yet, consider that you might be blocking or avoiding it; try to make space for it and invite it in. Try to *visualize* what the collapse of civilization will look like, and imagine what it will feel like. Use your imagination to explore what it will feel like while civilization is collapsing: No water coming from the tap; not knowing whether to stay or go. Mass starvations and migrations growing. If you can't imagine it for yourself, read Octavia Butler's *Parable of the Sower,* a novel set in 2025 in partially collapsed southern California. Or read *The End We Start From* by Megan Hunter. Their depictions of mass internal migration, horrendous flooding, and social breakdown may be the jolt you need.

Most psychological and sociological writing about climate fear focuses on the primitive human responses to crises — "fight, flight, or freeze" — or, alternatively, on the devastation of post-traumatic stress disorder (PTSD). This bleak portrayal of reactions has contributed to collective paralysis. While people can respond to fear by fighting, fleeing, being unable to react, or by later developing PTSD, these, as I've argued, are not our only options. We can also be inventive and collaborative in our response. We can use our fear to respond to crises with rationality, focus, moral purpose, and shocking efficacy.

We can transform our economy and society to beat back a global catastrophe. Indeed, we've done it before — and rapidly.

But to solve an emergency like the climate crisis, we need to collectively and immediately exit "normal mode" and abandon the gradual policy advocacies and enervated emotional states that accompany it. We need a collective awakening, on the scale of our response to a national attack. Together, we need to realize, in accordance with Wilson's theory of group selection, that we face an existential emergency, that we are in clear and imminent danger, and that we must immediately mobilize with everything we've got.

We can do this only when we enter into what I've called "emergency mode." In emergency mode, we can channel our fear to fight the threat facing us collectively and creatively. This is the opposite of panic mode, in which we either freeze or take flight. When we enter emergency mode, we respond to threats with thoughtfulness, planning, and coordination. Our senses are heightened; we feel a preternatural calm; we are able to process large amounts of information quickly to arrive at a flexible course of action.

When a society enters emergency mode, it mobilizes and works collectively to address and solve huge problems quickly. We know we can mount this response because we've done something similar before. By looking to history, we can see a path forward for maximal impact on the most pressing problem of our time. But to take this path, as I've argued, we need to be strong enough and self-compassionate enough to *feel* the threat and to *feel* our fear. As individuals and as a society, we need to use this fear to wake us up. You already know that it starts with you, looking inward — feeling the unbearable urgency of the crisis in your bones, taking responsibility for solving the climate emergency, committing to fighting for all life, and stepping up to turn that commitment into high-leverage action.

The goal of the emergency climate movement, and of TCM, is to lead the public out of normal mode and into emergency mode. Emergency mode describes how individuals and groups function optimally during an existential crisis. When individuals and groups enter emergency mode, they position themselves to achieve incredible feats through intensely focused motivation and collaboration. Individuals and groups enter emergency mode when they *accept* the reality of a life-threatening emergency and *reorient* by

- Adjusting their hierarchy of priorities so that tackling the emergency is the clear top priority.
- Deploying all available resources to solve the crisis.
- Seeking personal gratification and self-esteem enhancement through engagement with the emergency. People seek to "do their part" to solve the crisis and build their skills to contribute more effectively.

Emergency mode is a fundamental departure from a so-called normal mode of functioning. In normal mode, the individual or group feels relatively safe and secure. Immediate existential or major moral threats aren't recognized, either because there are none or because the vast majority of people are in denial.

We don't need to look very far back into history for examples of how entering emergency mode can solve pervasive, seemingly

	Normal Mode	Emergency Mode
Priorities	Many balanced priorities	Overriding priority: Solve the crisis
Resources	Distributed across priorities and saved for the future	Huge allocation of resources toward the solution
Focus	Distributed across priorities	Laser-like focus
Self-Esteem Source	Individual accomplishment	Contribution to the solution

intractable problems. During WWII, our whole country entered emergency mode. What the Allied powers accomplished during the 1940s is not only inspiring, it's a model for how we can collectively and effectively enter emergency mode today. We must look to our country's collective response to the threat of Axis power, to the specific methods by which we transformed our economy, shattered every production record to arm ourselves and our allies, and won WWII.

After years of bitter acrimony over the New Deal, WWII prompted conservative business titans and "New Dealer" government officials to join together, albeit in an uneasy alliance, to focus America's industrial might against the Nazis and Imperial Japan. Factories were rapidly converted from producing consumer goods to producing tanks, guns, bombs, and planes. The quick turnaround and staggering output shattered all historical records for war production.[93]

During this time, all hands were on deck, and everyone contributed what they could: Young men sacrificed their lives fighting for our country; women surged into factories to produce war material; Native American Code Talkers developed a system based on several indigenous languages to transmit secret messages for the allies;[94] and scientists and universities pumped out research on behalf of the war effort, leading to huge technological and intellectual breakthroughs. Fifteen million Americans, more than 10 percent of the population, relocated to find a war job, often across state lines[95] and more than 40 percent of vegetables were grown at home in Victory gardens.[96]

The response to WWII led to what I call "maximum intensity mobilization." This is when society as a whole enters emergency mode, responding to an emergency by collectively directing its energies to restructure its industrial economy and society immediately. Maximum intensity mobilization requires the support and involvement of the vast majority of citizens and the redirection of a very high proportion of available resources, and it impacts every part of society. Between 1943 and 1945, the United States spent more than 40 percent of its gross domestic product — almost half of the entire economy — on the war effort.[97] Maximum intensity mobilization is nothing less than a government-coordinated social and industrial revolution.

When an entire nation enters emergency mode, the results can be truly staggering. By mobilizing for total victory, the United States achieved goals it could never have reached in another way. When the United States entered WWII after the attack on Pearl Harbor in December 1941, President Franklin D. Roosevelt (FDR) laid out terrifically ambitious production targets for tanks, ships, guns, and airplanes. FDR set the goal of producing 60,000 planes in two years. People were deeply skeptical about whether such a feat could be accomplished. Yet, by 1944, the United States had produced 229,600 planes — more than three times the original goal.[98] In response to a cutoff of critical rubber supplies in Southeast Asia, the federal government launched a program that scaled up synthetic rubber production from less than 1 percent to about 70 percent of total U.S. production — a *100-fold* increase — in about four years.[99] In 1943, reclaimed rubber from citizen-coordinated scrap drives provided about 40 percent of the domestic rubber production.[100]

The United States also made huge advances in the sciences, in part by generously funding research of many kinds.[101] The first computer was invented during this time, as were plasma transfusions and sonar technology. The Manhattan Project successfully built the world's first atomic bomb in less than three years — a morally fraught but clearly stupendous feat of planning, cooperation, and scientific ingenuity. As an integral part of the mobilization during the multiyear emergency of WWII, the United States managed to maintain and, in some cases, expand, its basic systems, including infrastructure, education, health care, and childcare, and ensure that the basic needs of the civilian economy were met.

We do not need to have an overly rosy view of that time to appreciate the transformative effects of the United States' commitment. We cannot overlook the racist policies and attitudes — the military itself and many of the industrial mobilization jobs were segregated, and more than 100,000 Japanese Americans were interned. However, we can accept the painful reality of the United States' history of racism and still acknowledge the transformative potential of mobilization.

In fact, during WWII mobilization, major strides were made toward both racial and gender equality. FDR created the Fair Employment Practice Committee to investigate claims of discrimination in response to the fierce organizing of civil rights leader A. Philip Randolph and the March on Washington Movement.[102] The Double V Campaign advocated for victory for democracy and equality for African Americans overseas and at home.[103] Five million women joined the workforce for the first time,[104] and daycare centers were built in factories to provide childcare support.[105] Some factories also provided mothers with prepared dinners, so that they could work a full shift and still provide a hot meal for their children.[106]

A sense of national purpose and incredible energy suffused the entire country.[107] Because they were in emergency mode, citizens felt intensely motivated and made many sacrifices. They invested their cash savings in war bonds.[108] They tolerated a significant increase in taxes: The percentage of the population paying any income taxes jumped from 7 percent to 64 percent. Tax increases were focused on the rich. The top marginal income tax rate on the highest earners reached 88 percent in 1942 and a record 94 percent in 1944 on income above $200,000 — the equivalent of about $2.85 million in today's dollars. A tax on excess corporate profits provided about 25 percent of government revenues during the war.[109]

Historian David Kaiser describes the emergency-mode mindset that pervaded the country during WWII: "At no time in

American history have they shown more willingness to make financial sacrifices to meet common necessities, largely because they agreed with their president that the survival of civilization was at stake."[110]

The federal government instituted a sweeping rationing program to ensure fair distribution of scarce resources on the home front and to share the sacrifices equitably. Gasoline, coffee, butter, tires, fuel oil, shoes, meat, cheese, and sugar were rationed. American historian Doris Kearns Goodwin describes the impact of these measures:

> By and large, American housewives accepted the system of rationing cheerfully...Citizens learned to walk again. In the months that followed, car pools multiplied, milk deliveries were cut to every other day, and auto deaths fell dramatically. Parties at homes and nightclubs generally broke up before midnight so that people could catch the last bus home. All in all, pleasures became simpler and plainer as people spent more time going to the movies, entertaining at home, playing cards, doing crossword puzzles, talking with friends, and reading.[111]

This is a powerful example of what can happen when a government and society enters emergency mode and commits to maximum intensity mobilization. It differs markedly from a business-as-usual or normal "political paralysis" mode, as David Spratt and Philip Sutton outline in their groundbreaking book, *Climate Code Red: The Case for Emergency Action:*[112]

Maximum intensity mobilization is an economic approach that directs the collective force of industry away from consumerism and toward a singular national purpose. Profit-seeking activities are channeled toward the national mission and are tightly controlled if excessive. Emergency mobilization is characterized by large-scale deficit spending; sweeping command-and-control regulations; a high degree of citizen participation; increased taxation, especially of the wealthy, to control inflation and raise

Normal "political paralysis" mode:	Emergency mode:
Crises are constrained within a business-as-usual mode	*Societies engage productively with crises — not in panic mode*
Spin, denial, and "politics as usual"	The situation is assessed with brutal honesty
No perceived urgent threat	Immediate or looming threat to life, health, property, or environment
Problem not yet serious	High probability of escalation beyond control if immediate action is not taken
Time of response is not important	Speed of response is crucial
One of many issues	Highest priority
Labor market	Emergency project teams, labor planning
Budgetary "restraint"	Devote all available/necessary resources; borrow heavily if necessary
Community and markets function as usual	Non-essential functions and consumption may be curtailed or rationed
Slow rate of change due to systemic inertia	Rapid transition; scaling up
Market needs dominate thinking and response choices	Planning and fostering innovation
Targets and goals determined by political tradeoffs	Targets and goals not compromised
Culture of compromise	Failure is not an option
Lack of political and national leadership; adversarial politics	Heroic leadership; bipartisanship

revenues for the national project; and strong government control over the allocation of raw materials and basic goods. And although corporations can play a constructive role in implementing mobilization, they do not drive the change process.

This type of all-in, full-scale mobilization is the only approach comprehensive enough to resolve a crisis on the scale of the climate and ecological emergency. Its implementation also has many other benefits. On the WWII home front, for example, mobilization drastically reduced income inequality, helped ensure full employment, and reinforced the importance of cultivating the participation of every citizen.[113]

The maximum intensity mobilization mentality is the opposite of consumerism. People ask not "What do I want to buy?" but rather "How can I contribute?" This mentality can foster an increase in societal trust and a stronger sense of national purpose.[114] For those who accept the need to rapidly — not gradually — convert an economy to a new purpose, maximum intensity mobilization is the most effective, egalitarian, healthy, and sensible approach.

TCM has spent years considering how the example of homefront mobilization during WWII could be applied to the current climate crisis. Ezra Silk, TCM's cofounder and director of strategy and policy, wrote The Climate Mobilization's *Victory Plan* to demonstrate what a 10-year national maximum intensity mobilization plan to address the climate and ecological crisis could actually look like. Our *Victory Plan* outlines an emergency mobilization that aims to:

- Create a zero-emissions greenhouse gas economy in 10 years and draw down excess greenhouse gases until a safe climate is restored.
- Shrink the ecological footprint of the global economy from 1.7 planets per year to approximately half a planet per year.

- Halt the sixth mass extinction by creating an interconnected global wildlife corridor system covering half the earth's land surface and oceans.
- Create a society and global economy that works for everyone, including a healthy and stable global environment, healthy food, clean air and clean drinking water, life-affirming work at a living wage, medical care, housing, and full democratic participation in government and at the workplace.

Our *Victory Plan* includes policy recommendations, such as an immediate ban on all new fossil fuel infrastructure and a 10-year period in which the federal government phases out the oil, coal, and gas industries, likely through a rapidly declining cap on fossil-fuel extraction and imports. It details massive energy conservation and renewable energy investments, as well as a new "super-smart" electricity grid. It calls for banning factory farms and pesticides, and for launching a national program supporting regenerative agriculture and more plant-based diets. It bans single-use plastic, fostering a circular economy in which nothing is wasted. It includes reforestation and rewilding projects to restore biodiversity. All hands will be on deck — the federal government will guarantee a job to anyone who wants to take part in the mobilization. Others can participate by cutting their home energy use, biking instead of driving, growing food in Victory gardens, and participating in community energy conservation efforts. Everyone will have a role to play.

The vision articulated in *Victory Plan* has already inspired others to raise their expectations of what's possible. It has influenced the policy vision proposed by Representative Alexandria Ocasio-Cortez and the Justice Democrats in their Green New Deal legislation, which calls for a 10-year WWII-scale mobilization and describes a range of programs to ensure racial, gender, social, economic, and environmental justice. These measures would also provide tremendous stability for displaced workers and all Americans during incredibly rapidly changing times. TCM calls for maximum intensity mobilization because we know that there is no faster, more powerful way to transform our economy. The Greatest Generation mobilized to beat back an existential threat, and we can, too. Once we have fully entered emergency mode and mobilized, we will be optimally positioned to shatter every record and expectation, create new discoveries, restore a safe climate and threatened ecosystems, and create a society based on meeting human needs and protecting all life.

But *how* we make this happen is the defining question of our moment. It's the question that should keep everyone who is living in climate truth up at night. The call for WWII-scale climate mobilization has been made by Senator Bernie Sanders, economist Joseph Stiglitz,[115] *The New York Times* columnist Thomas Friedman,[116] environmental activist Bill McKibben,[117] and many more. But *knowing* or even *arguing for* what is necessary doesn't necessarily ensure it will happen.

To solve an emergency like a world war or the climate crisis, we need to exit normal mode collectively. We each need to enter emergency mode ourselves and then communicate our choice to others. Because, as I've already argued, the way we respond to threats — whether by practicing denial and remaining in normal mode or by facing our fears and entering emergency mode — is *highly contagious*. We can initiate collective behavior by first transforming as individuals, then by leading others.

Let's take another look at how the United States entered emergency mode during WWII. Before the attack on Pearl Harbor, most of the public and their representatives in Congress had minimized the threat of the Axis powers. Even as Germany swept through Europe, isolationism dominated American politics. The prospect of fighting a war was extremely unpopular. World War I had left people disillusioned by war, and the Great Depression was still blighting the country. There was a widespread feeling that the trouble in Europe should stay in Europe.[118] However, there were leaders, including FDR, who understood the danger and were preparing the military, and the public, for war.

The United States entered emergency mode in 1941, after the surprise attack on Pearl Harbor. That event triggered an overwhelming national consensus in favor of war. Congress immediately and unanimously voted to authorize war. As Goodwin writes:

> Isolationism collapsed overnight. "American soil has been treacherously attacked by Japan," former President Herbert Hoover stated. "Our decision is clear. It is forced upon us. We must fight with everything we have." Senator Arthur Vandenberg of Michigan, who had struggled long and hard against American involvement in the war, phoned the White House to tell the president that he "would support him without reservation." Even Representative Hamilton Fish of New York, one of Roosevelt's severest critics, urged the American people "to present a united front in support of the President." [119]

According to psychologist Daniel Gilbert, humans are wired for a reflexive response to threats that are "intentional, immoral, imminent, and instantaneous." [120] The attack on Pearl Harbor was marked by all of these characteristics. The shock and feeling of betrayal created a recognizable and collective feeling of fear and vulnerability, as well as anger, and a fierce desire to *respond* by fighting back. Plus, the groundwork had been laid by the preparedness campaign led by President Roosevelt and a group of forward-thinking government and military officials, as well as business leaders who had already started the rearmament processes.

A surprise attack isn't the only way to provoke collective entry into emergency mode. Social movements have successfully illustrated the efficacy of emergency mode, too. AIDS Coalition to Unleash Power (ACT UP), an activist organization that is working to end the AIDS pandemic, shows how a citizen group can actively work to change a government response and provoke comprehensive change. Through its actions, ACT UP made the federal government treat the AIDS epidemic like the public health emergency it is.

In the 1980s, HIV, the virus that causes AIDS, was spreading at horrifying speed and decimating gay communities in New York, San Francisco, and other large cities. The government, as a result of pervasive homophobia, was giving HIV victims no help, and neglecting research and treatment. The government's failure to act destroyed whole communities.

It took now-iconic AIDS activist Larry Kramer to channel the terror and anger of the gay community and their allies into a social movement capable of demanding an appropriate response from the government and other institutions. Although Kramer cofounded the Gay Men's Health Crisis (GMHC), he broke with the group over disagreements about strategy and tactics in responding to the AIDS crisis. According to Kramer, the GMHC did not enter emergency mode. Instead, it continued to seek solutions through business-as-usual channels, holding meetings with government officials to ask for help. These strategies were not working. Kramer criticized GMHC, arguing that it was so focused on claiming mainstream status that it was helping people die rather than fighting to protect the living. In response, Kramer cofounded ACT UP — a nonviolent but militant organization that viewed AIDS as an existential threat. ACT UP did not politely ask the government for help: It *demanded* emergency action.[121]

Kramer knew he was fighting for his own life and for the lives of his friends. He had no interest in "business as usual." He wanted the government to act on AIDS *now* — to research the illness, find treatments, treat the sick, and prevent transmission. Kramer treated AIDS with deadly seriousness, using fear and inflammatory rhetoric to provoke people to action. In fact, he exhorted his would-be supporters to feel as much fear as possible — telling crowds of gay men that if they didn't fight back, they would soon be dead. Kramer referred to AIDS repeatedly as a "plague" and to the politicians who ignored it as "Nazis" and "murderers."[122]

The symbol of ACT UP, a pink triangle, is a reference to the genocide of gay men during the Holocaust. Its slogan, "Silence = Death" referred not only to the government and media's silence on AIDS but to the cultural silence around homosexuality. At the time, many gay people were closeted, hoping to avoid discrimination and dehumanization from a homophobic culture. Kramer's solicitation of fear and his bracing language and symbology were tactical: He was not just inviting other gay men to join him in emergency mode and to focus intensely on solving the crisis; he was emphatically arguing that there was no alternative.

The silence around homosexuality, with most people keeping their sexual orientation at least partially private, posed a huge problem for the movement. It contributed to the ignorance about AIDS. Gay people, including gay government workers, gay researchers, gay doctors, and others, could not work together with maximum impact or collectively communicate the emergency to the public while still in the closet. The public assumed that if they weren't hearing much about a health crisis, there must not be one. Kramer spoke to this in his landmark essay, "1,112 and Counting":

> Why isn't every gay man in this city so scared shitless that he is screaming for action? Does every gay man in New York want to die?...I am sick of closeted gay doctors who won't come out to help us...I am sick of closeted gays. It's 1983 already, guys, when are you going to come out? By 1984, you could be dead. Every gay man who is unable to come

Credit: ACT UP

> forward now and fight to save his own life is truly helping to kill the rest of us. There is only one thing that's going to save some of us, and this is numbers and pressure and our being perceived as united and a threat. As more and more of my friends die, I have less and less sympathy for men who are afraid their mommies will find out or afraid their bosses will find out or afraid their fellow doctors or professional associates will find out. Unless we can generate, visibly, numbers, masses, we are going to die.[123]

This push to come out of the closet without shame was prominent and powerful throughout the AIDS movement. How many people came out in response to the AIDS crisis? How many individual conversations were had among families, among friends, among colleagues? Perhaps millions. Learning that people they loved and respected were gay and in danger had a profound impact on many people. Homosexuality and AIDS could no longer be an abstract phenomenon that affected "other" people. Breaking the silence brought the crisis home.[124]

ACT UP also regularly held creative, disruptive protests to demand that the government launch a crisis response to the AIDS epidemic. They channeled their grief and terror into effective action. In *This is an Uprising*, author-activists Mark and Paul Engler describe the group's hard-hitting approaches: chaining themselves to buildings; blocking FDA offices and papering the offices with posters of bloody handprints; stopping traffic in high-profile places like the Golden Gate Bridge; interrupting CBS's nightly newscast with on-air protests; and spreading ashes of dead loved ones on the White House lawn.[125]

Members of ACT UP recognized AIDS as an existential threat and entered emergency mode, channeling their emotion-fueled energy, focus, creativity, and resources toward responding to the emergency. ACT UP became the center of political and social life for many of its members.[126] ACT UP didn't focus on using science or data to convince people that AIDS was a crisis. Instead, they *demonstrated* their understanding of HIV as an emergency through their actions. They obeyed the axiom of effective communication: "Show, don't tell." People who saw ACT UP's actions got the message loud and clear that something must be terribly wrong. Many who saw ACT UP in action also entered emergency mode and joined the fight — contributing their skills and resources, mobilizing their networks, and joining protests.[127]

Of course, entering emergency mode doesn't mean you *only* undertake disruptive protests. Policy advocacy, research, and education are also critically important in creating social change. ACT UP's groundbreaking Treatment and Data Committee took on the task of becoming experts in the biology of HIV/AIDS — seeking to understand the virus and various treatment options. ACT UP created a glossary of AIDS treatment terms to pass out at meetings. The group also produced and advocated through "A National AIDS Treatment Research Agenda," which detailed ACT UP's specific demands for the drugs that should be developed and the ways the research should unfold.[128]

ACT UP illustrates that a small group of people can enter emergency mode and mobilize to become a formidable group of people in emergency mode that can lead the government itself into emergency mode.[129] In part as a consequence of ACT UP, the government mobilized its research and public health apparatus. AIDS patients won the right to participate in every phase of the drug development process. They won major funding for research, which led to the discovery and deployment of life-saving drugs. In fact, ACT UP helped to change the way pharmaceutical drugs are researched and developed.[130]

Although ACT UP's success was incomplete — AIDS became a pandemic, and one million people still die from AIDS every year — ACT UP mobilized an incredibly effective response to the HIV crisis in the United States. Its success also laid the groundwork for mainstream acceptance of homosexuality in America and gave energy to the continuing struggles for gay rights and equality everywhere. It took contributions from researchers, doctors, nurses, policymakers, public health officials, journalists, government officials, and more — who worked tirelessly for more than 30 years — to create today's conditions, where more than 23 million people globally are receiving highly effective antiretroviral treatment.[131,132] ACT UP was able to instigate this by leading individuals and institutions like local, state, and national governments; the National Institutes of Health; hospitals; and universities to treat HIV/AIDS like the crisis it is.

The examples of WWII and ACT UP provide a picture of what entering emergency mode and initiating maximum intensity mobilization could look like. Yet the picture is only partially relevant. American entry into the emergency mode of WWII was the result of preparation from leadership *and* a universally shocking trigger. A foreign attack is the most common historical trigger for a society to enter emergency mode. An outside threat provokes people to freeze, fight, or flee. After Pearl Harbor, the United States entered emergency mode and *fought* with all available resources; the entire country focused on an overriding national goal.

Today, thanks to the lies of corporations, and the complicit silence and understatement of our government, media, institutions, and the mainstream climate movement regarding the climate crisis, we haven't had a universally shocking trigger to galvanize support. Neither have fearsome scientific reports, superstorms, mudslides, or wildfires yet proven to be our "Pearl Harbor Moment." When we compare FDR's leadership in the run-up to WWII to the Trump administration's total failure of leadership, we can only shudder. Roosevelt was well respected, although hated by much of the business class, and had already led the country through the Depression with the New Deal, which addressed but did not cure, the nation's economic woes.[133] In general, Americans had a basic sense that their government had a strong leader who was able to take on critical challenges.[134] In marked contrast, the Trump administration is disrespected, incompetent, corrupt, and demonstrably against acknowledging the climate crisis. The Trump administration is fundamentally uninterested in engaging in mobilization and would be woefully unequipped to lead one, even if it did wake up to the existential risk.[135] Americans understand this. Data from The Pew Center shows that the public's general trust in government is near historic lows, with only 17 percent of Americans believing that the federal government does the right thing nearly all or most of the time.[136]

Today is so vastly different from the 1940s that to spark a Pearl Harbor-level awakening, we must pursue the ACT UP model. Once again, *silence equals death*. It's time to get loud. We must all recognize that the climate crisis is a clear and present danger. We must spread the truth about the climate emergency and the need for mobilization, through word and deed. As Extinction Rebellion puts it, "Tell the truth and act like that truth is real."

We have almost everything we need to draw down excess emissions and eliminate greenhouse gas emissions in less than 10 years. Renewable energy has made tremendous technological advances in recent years and is rapidly becoming cheaper than fossil fuels. With enough investment in infrastructure and energy conservation, renewable energy could power the entire country, indeed the entire world, by 2030.[137] High-speed rail plans have been drafted.[138] Permaculture techniques can be applied to farming practices to sequester more carbon in the soil.[139]

We must deploy thousands of shovel-ready solutions at scale. To enact these solutions, though, we need a federally coordinated program of public planning, global cooperation, massive public investment, forceful regulations, economic controls, public-private partnerships, and full societal participation.

We can do all of these things if we enter emergency mode. All we're missing is a collective awakening to summon tremendous political will. We need millions of people to feel their pain and devote themselves to the cause of building power to demand action. We need a national, and then global, consensus that we are *in danger and need mobilization to protect ourselves.*

That's why I speak to the importance of facing our fears: We need to feel scared enough to transform, whatever it takes. That's also why I argue that individuals who understand the climate emergency and mobilization imperative need to lead this transformation. We need to act as messengers and prophets. We need to be brave — even heroic — in our ability to tell the truth. We need to get talkative and loud. Like Kramer and his supporters in ACT UP, we must spread the truth as clearly, vociferously, and in the most attention-grabbing ways we can. This means art, slogans, mass meetings, teach-ins, public demonstrations, and protests. We must not stay "closeted," appear to believe that everything is fine, or continue with life as usual. Instead, we need to "come out" — to our friends, family, neighbors, fellow climate activists, and the public. We must be adamant: Gradualism is not enough. We are in danger; we are in emergency mode; and we will do everything we can to initiate a WWII-scale climate mobilization — the only thing that can protect us.

Questions for Reflection and Discussion

- What did your family members do during WWII? What are their memories and stories from that time?
- Have any of your family or friends been involved in ACT UP or the AIDS movement or other social movements?
- Did you ever have to fight for your life and for the lives of your friends? What was your experience?
- Have you ever waged a battle against silence and lies using the truth as your core weapon? What worked, and what failed?

STEP FIVE:

Join the Climate Emergency Movement

ARE YOU READY? Have you faced climate truth and mourned your losses? Are you building emotional muscle — confronting your defenses and experiencing fear and other uncomfortable feelings? Are you convinced that nothing matters more than solving the climate catastrophe? Have you learned from the examples of history? Are you ready to do whatever you can to protect humanity and the natural world from calamity? If so, welcome to the team — the climate emergency movement. We're a diverse group made up of individuals across the nation and the world. We demand that governments treat the climate like the existential emergency it is and initiate maximum intensity mobilization to restore a safe climate. We are a movement transforming fear and grief into real and meaningful action. We are a movement made up of different organizations all over the world, including the school climate strikers, Sunrise Movement, Extinction Rebellion, the Justice Democrats, and, of course, TCM. We are nature defending itself; we have joined up with the force on Earth that wants to live, and we draw tremendous strength from that.

Although our movement has been building for a long time, the shift from incremental goals toward a real fight for our species' survival has been so radical and so recent that it is almost hard to grasp the change in tone and seriousness. In some assessments, that shift began in 2016 at Standing Rock, where the indigenous Water Protectors showed the country what heroism looks like, withstanding months of abuse from police and private security forces while continuing their nonviolent, highly spiritual direct action of blocking the Dakota Access Pipeline. Standing

Rock Lakota Nation member Floris White Bull describes Earth's vital life force calling her to become a Water Protector:

> I've been woken
> by the spirit inside that
> demanded I open my eyes
> and see the world around me.
> Seeing that my children's future
> was in peril. See that my life couldn't
> wait and slumber anymore. See that I was
> honored to be among those who are awake.
> To be alive at this point in time is to see the rising
> of the Oceti Sakowin. To see the gathering of nations
> and beyond that, the gathering of all races and all
> faiths.
> Will you wake up and dream with us?
> Will you join our dream. Will you join us?[140]

White Bull's call inspired others, including politician Ocasio-Cortez, who cites her time at Standing Rock as critical to her decision to run for Congress.

The climate emergency message has broken through to the wider public only recently. Until late 2018, TCM was the only national group calling for a 10-year race to zero CO_2 emissions; Sunrise Movement was almost unknown outside activist circles; Extinction Rebellion was new and had yet to make a name for itself; and the primary national climate policy advocated by champions like Bernie Sanders and 350.org still advocated a gradual shift to renewable energy by 2050 as their primary national climate policy.[141]

Fast-forward one year, and we are in a "moment of the whirlwind."[142] Thanks in part to the prefiguring work of TCM, organizations all over the world have begun to step up and coordinate climate emergency efforts. Sunrise Movement, for example, has mobilized a movement of young people in the United States who have been occupying the offices of elected officials who refuse to act with urgency: Millions of student school strikers are demanding an emergency response. Extinction Rebellion demands that governments tell the truth and declare a Climate Emergency, implement zero emissions by 2025, and govern the mobilization via citizen assemblies.

TCM helped pioneer the Climate Emergency campaign, which has used Declarations of Climate Emergency as a highly successful campaign tactic. The Climate Emergency campaign officially started in the city of Darebin, Australia, which passed the first Declaration of Climate Emergency in December 2016.[143] In November 2017, TCM's local chapter in Hoboken, New Jersey, persuaded the city government to become the first U.S. city to declare a climate emergency resolution. Working in a coalition with Extinction Rebellion, Greenpeace, international Green Parties, and on-the-ground leaders, TCM has helped to spread this campaign to more than 1,300 local governments around the world, representing more than 800 million people.[144] By the time this book prints, there will surely be hundreds, if not thousands, more.

When cities or governing bodies declare a climate emergency, they communicate a critical, too-often-unspoken truth. They are shifting their paradigm and positioning themselves to take action. TCM recommends that cities that declare climate emergencies "ban, plan, and expand" — ban new fossil-fuel infrastructure and other destructive practices; start a planning process with community participation about how to rapidly decarbonize; and expand climate emergency mobilization, promoting it to the public, to other cities, and to higher levels of government.

It is hard to overstate what a monumental shift we are experiencing. You can and must take part in this historic movement. By joining the climate emergency movement, you choose the heroic road over the easy road; you choose mission over self.

Have you imagined a heroic version of yourself? Have you ever dreamt that you — yes, *you* — could actually help save the world? Try to get in touch with this feeling. Let yourself daydream about answering the call to heroism. Fight off thoughts that tell you it's egotistical or silly to think that *you* can be a hero and help save humanity. Let yourself *want* that heroism, and then let yourself strive for it. *Now* is the time. The climate emergency movement is here, and it needs you. Its success depends entirely on whether people like you step up and join in.

What role you play in the movement is a complex and important choice, although I encourage you to jump in, get started, and approach the question in an ongoing way. Be thoughtful, flexible, and strategic about your role as you go, always reaching toward the highest impact.

Start by telling the truth, loudly and all the time. This is the one mode of engagement that I recommend for everyone. Start with your friends and family, and expand from there. A recent study from Yale University confirmed the impact of this approach, finding that "discussing global warming with friends and family leads people to learn influential facts, such as the scientific consensus that human-caused global warming is happening. In turn, stronger perceptions of scientific agreement increase beliefs that climate change is happening and human-caused, and increase concern about climate change."[145]

If we are silent, we are powerless. Even today, despite the moment of whirlwind we inhabit, most people remain silent: Polling from The Yale Program on Climate Change Communication found that only 8 percent of Americans hear people they know talk about climate change at least once a week, and only 15 percent once a month.[146] Thirty-nine percent of Americans hear about climate change from someone they know only "several times a year" or less, and 24 percent of Americans have never heard anyone they know talk about climate change. The majority of Americans are either "concerned" (30%) or "alarmed" (29%).[147] This is pluralistic ignorance in action. We assume we are alone in our fear and grief, and we stay silent.

It's past time to let your friends and family know how you feel about the climate emergency. Perhaps you worry about the social awkwardness of bringing up the climate emergency — and it may be awkward at first. But keep in mind that many of your friends and family are also worried, and they will be relieved and appreciative when you bring up your feelings, especially if you can offer them support and guidance. Be personal; be emotional; be authentic and empathetic. Hear people out and make them feel listened to. Talk about this book — sharing it is a great way to broach your conversations. Try saying, "I know we've shared our concerns about the climate crisis, and this book has really helped me. Would you be willing to take a look and tell me what you think?"

Once you're comfortable talking about the climate emergency with your friends and family, begin to talk with others. Set a goal to talk about the climate emergency and the need for climate mobilization once a day; then, build up to talking about it more than once a day. Consider showcasing the climate emergency message on t-shirts, hats, pins, or bags. This is another way you can communicate your affiliation with the movement and invite potential conversations with others.

Start having these conversations online, too. You can talk about the climate emergency and the need for mobilization on social media and, depending on your access, on email lists, blogs, or in mainstream publications. How many followers do you have on Twitter, Instagram, and Facebook? Do you use Reddit or Medium or make YouTube videos? You may be able to make a significant impact by spreading climate truth through those channels.

Social movements have long utilized cutting-edge communications technologies — which have not yet been controlled and co-opted by the powerful — to fight denial and spread their messages. In his time, Martin Luther used the revolutionary potential of a new technology — the printing press — that led to the Protestant Reformation. Back in 1518, printing hundreds of copies of your political arguments and distributing them was an innovation, and a very effective one.[148]

Martin Luther King, Jr. used television as a tool to broadcast the violence of segregation into millions of American homes.[149] Civil disobedience created hundreds of dramatic, suspenseful scenes, such as confrontations during lunch-counter sit-ins. The public was captivated. What would happen? How would this turn out? How would the owner and waitstaff respond to this protest? How would the protesters respond to abuse? Would law enforcement get involved? Would there be violence? Would people die? For many, these scenes unfolding on the news night after night were a spear in denial's heart. Think the system isn't so bad? Look what happens to those who challenge it. The brutality and oppression inherent in the Jim Crow system, as well as the dignity and humanity of African Americans, were brought by television into Americans' living rooms.[150]

For climate warriors, the Internet is our cutting-edge technology. Social media offers the possibility for social movements to burst into the forefront of the public consciousness. The #MeToo movement offers a recent example. When women began sharing their individual stories of mistreatment on social media, it caused a lightning-fast transformation of the political landscape. An oppressive silence was broken, and powerful men finally began to be held accountable for sexual harassment. The #MeToo movement illustrates, once again, the centrality of integrating the intellectual and emotional, the personal and political. Women sharing their stories of sexual harassment en masse made many men finally realize the huge scope of this problem, and made institutions realize that they needed to act.

Imagine if a #NoFuture movement started about the climate emergency, with first a handful, and then hundreds, thousands, and even millions of people sharing their personal reactions to the climate crisis — expressing their fears and revealing their dashed hopes for their futures. An online movement like this on the scale of #MeToo — combined with street action — might break the dam of denial and help provoke a great awakening.

To prepare for your role as a loud and talkative truth-teller, ask yourself how comfortable you are talking about the climate emergency. If you find that you are very uncomfortable, identify what makes you feel that way. Then practice navigating these discomforts by role-playing with friends. Later, when engaging in low-risk conversations with friends and family, ask for their feedback. Did your words have an impact? What do they need to hear to provoke them to action? Don't feel discouraged if the feedback you hear isn't entirely positive, and don't weigh any one person's comments too heavily. Different people respond to different approaches. Try things out. Expand your comfort zone.

There are many options for your next step. Consider joining a climate emergency group and leading your city government to declare a climate emergency. You can join Extinction Rebellion, and plan and execute nonviolent direct actions. You can organize with Sunrise Movement, the Justice Democrats, or one of the other groups supporting the Green New Deal. You can volunteer for TCM as a researcher, publicist, or social media expert. You can join a school climate strike, organize within your workplace or church, or canvas for a climate emergency candidate in a primary. There are many ways to offer support. You could, for instance, provide childcare for people attending organizing meetings. Or you could provide translation services, cook for protesters, copyedit press releases, compile newsletters, or keep spreadsheets updated. The most important thing is to get started.

The material that follows will help you assess and practically place yourself as a climate warrior among other passionate, dedicated, and talented heroes within the climate emergency movement. Once you've recognized and accepted that you must take drastic measures to save our planet, the question "What can I do about it?" is *the most important question each of us can answer.* This isn't easy. You may have already tried to act. You may feel that it simply isn't always clear *how* to convert fear, grief, and feelings of responsibility into effective action.

In my work as the founder of TCM and with my experience as a clinical psychologist, I work every day to guide the commitment of people who are ready to contribute to the climate emergency movement. It's very complex stuff! Finding your place in the climate emergency movement is as multidimensional and personal as choosing a career path. Ultimately, only you can make the decision. However, by helping you identify your strengths and abilities, I can help match you to the movement for maximum effectiveness. Our goal is to find the place where your abilities can be put to their highest and best use, where you see your impact, and where you are engaged and challenged and pushing yourself, yet avoiding burn-out.

Below I describe several opportunities for engagement. Here, you'll seriously consider your availability, skills, networks, and resources. Keep in mind that no one has strengths in every area; it's just as important to the movement for you to identify areas for growth and development. When you've answered the questions, ask a friend or colleague for their thoughts.

The best volunteers have assessed their schedules, determined the number of hours they can commit to the cause, and joined the movement in a capacity that matches their passion and their availability. They are flexible; they are responsive; they are reliable; they take initiative. If you've read this far, I may well be talking about you — a thoughtful, committed, flexible, reliable, empowered climate warrior who can make a difference in an organization like TCM and, by extension, the world.

However, it may take some flexibility and persistence. It can be frustrating for volunteers to feel their skills are not well used, but keep in mind that climate emergency organizations are usually small and budget-constrained. Onboarding and managing volunteers can be quite time-intensive. Organizations can make great use of reliable volunteers who are flexible enough to take on any task that needs doing. Consider your role as a supporting one, at least in the beginning. Work to support others in what they are already working on; get to know the lay of the land; and grow into a leadership role.

To determine your role in the climate emergency movement, ask yourself how many hours per week you can commit to climate work. Aim high. If your first reaction is that you don't have much time to devote, think again about that reflexive response. You wouldn't be too busy to fight a raging fire, would you? Your time is valuable and is no doubt limited, but push yourself to devote at least seven hours a week to the climate emergency movement. While a volunteer with 25 hours a week can develop and manage an entire outreach campaign, a volunteer devoting seven hours a week can send recruitment text messages.

Obviously, the more time, the better. As the director of a startup organization that relies on volunteer time, I can't overstate the importance of volunteers who can dedicate 25 hours a week or more; it's just so much easier to coordinate people and get things done with people who are available throughout the day. Many volunteers and organizers sacrifice major comforts to give their time to a cause. In fact, many people at TCM have left high-powered jobs, moved in with their parents, and reined in their expenses to facilitate their round-the-clock climate work. If we are going to protect humanity and the natural world, some of us are going to need to go "all-in, for all life." Some of us must take responsibility for solving the climate crisis. Some of us must decide to sacrifice our ambitions and comforts. To determine if you are now or can be in this position, consider the following:

- What is the greatest hourly and weekly commitment you can make?
    - For six months
    - For one year
    - For three years
    - Until the next election?
- Could you reduce your hours at work or your time spent on other activities to allow for more climate work?
- Do you have savings or assets to cash in to support more climate work?
- Do you have a partner who is willing to support your ambition to work full time to save the world?
- Can you move in with family or roommates to cut your housing expenses?
- Can you ask the people who love and respect you to help support you to take a "climate year"?

Join a Local Climate Emergency Group. If you have four or more hours a week, consider joining a local chapter of a climate emergency organization. Is there a chapter of Extinction Rebellion, Sunrise Movement, TCM, or a school climate strike group in your area? Is there a chapter of 350.org, the Sierra Club, or another national organization that is working on (or is interested in working on) a Climate Emergency campaign? Are there local climate or environmental justice organizations that are working on a Climate Emergency campaign or that might be interested in doing so?

Working with local groups allows you to start walking the path of a climate warrior and leveraging a relatively limited amount of time. You can contribute to a group's ongoing efforts by making signs, flags, and other protest art; taking notes or offering child-care at meetings; cooking for protesters; offering bail support; holding gatherings in your home; and more. If you keep showing up, the opportunities to help will increase. Answering these questions can help you determine your best fit for a local chapter:

- What existing climate-emergency organizations are available in your area?
  - If you are a youth, consider Sunrise Movement.
  - If you're the parent of school-aged children, consider an organization that works to organize school climate strikes or works with students in other ways.
  - If you have a background in policy, lobbying, or community organizing, consider joining a politically oriented organization like TCM or an organization involved in Climate Emergency Declaration campaigns.
  - If you are willing to get arrested, give special consideration to Extinction Rebellion, which puts arrestable actions at the core of its strategy.

As you get to know your local climate emergency movement organizations, think critically about them: Are they firmly committed to telling the truth about the climate emergency and advocating for solutions that could actually work? Are they accomplishing their goals? Are they friendly and welcoming? Do they seem to have momentum and energy? Do they have a plan for the next three months, six months, and year? Are they putting your talents to good use? Can you grow in your role in this organization? If an organization isn't passing these criteria, reconsider your affiliation.

Declare a Climate Emergency. If you have seven hours or more a week and are already affiliated with an organization, whether it is a professional organization, a Rotary Club, a church, or a PTA, and you have strong leadership, communication, or social skills, commit to getting your organization to declare a climate emergency and take on campaigning for an emergency response. This kind of leadership — in communities and institutions you already know and are a part of — can be extremely important. Because trust and relationships are critical in accomplishing transformative change, bringing climate truth to an organization in which you already have relationships and have built trust allows you to accomplish more, quickly.

Assess your organization's leadership and membership, and determine who will be able to understand climate truth and who might want to get the organization more engaged. Building a small contingent within an organization is more powerful than acting as a lone champion (although "lone champion" is where many people start). Once you've built a small group of supporters, determine a strategy for moving your organization into emergency mode. Your strategy should include educating the organization's membership and leadership about the climate emergency, either through one-on-one meetings or larger presentations; getting a declaration of climate emergency onto the agenda of an upcoming board or membership meeting; and thinking strategically about how *this* organization can propel the movement forward.

To identify ideal organizations and determine if you are already positioned to work effectively with organizations in which you have influence, answer these questions:

- With which organizations are you currently involved?
- Are you a decision-maker in one of these organizations, or can you become one? If not, how much influence do you have in these organizations?

- Do you feel confident in your communication and leadership abilities? Is this a challenge you feel ready to take on?
- Who in these organizations might team up with you?
- Who are the most respected people in the organization, and how can you persuade them of the importance of addressing the climate emergency?

Become a Community Organizer. If you have 10 or more hours a week, you can become a community organizer and leader. Joining and supporting a local chapter of a climate emergency organization is a great way to get started, and becoming an organizer is the next step. Consider taking an active role in expanding Extinction Rebellion, the Sunrise Movement, or TCM. Community organizing is a deep commitment, but with preparation, delegation, and persistence, you can help these existing organizations grow and succeed. Each of these organizations offers organizer support — including guidelines, materials you can use, check-in phone calls, and more — so you won't have to start from scratch. If you are launching a new group, prepare for a more significant time commitment in the initial months of its startup. If you spend that time looking for others who can share the leadership responsibilities, you will be able to keep the time commitment manageable.

Your work in community organizing will involve recruiting, orientation, teaching, planning, coordinating, and other leadership functions. Prepare by heeding the words of brilliant organizer and writer Jane McAlevey: "Organizing isn't rocket science, but it's a serious skill and craft." [151] Learn the skills and theory behind the craft with *Raising Expectations (and Raising Hell)* by McAlevey, *Rules for Revolutionaries* by Zack Exley and Becky Bond,[152] and *Organizing for Social Change* by Kim Bobo, Jackie Kendall, and Steve Max.[153] To go deeper, consider an online or in-person training through Momentum, Movement School, or

Midwest Academy. Finally, seek mentorship from people who are already involved in organizing.

Your job will not be merely to execute a successful local Climate Emergency campaign; it will also be to recruit and develop new leaders. The heart of organizing is to lead and teach, and your job is to bring more people to the movement. You are not only spreading the climate emergency message but also teaching a set of skills that people can use to turn that message into power. Developing new leaders and organizers also means you can delegate responsibilities and have more flexibility yourself.

As a community organizer, a key part of your role is to help people see the truth and join the movement. You can help them consider *their* best ways of engagement. For example, you might ask an interested friend, "I need someone to make phone calls every week, reminding people to come to meetings. Could you do that?" To someone else, you might say, "Well, you have a law degree, so you could offer an organization legal advice, but you also have organizing experience...I wonder if you would like to organize within a bar association?" Helping people imagine themselves in different roles and get started is just as important as anything else you will do for humanity and the living world. When you undertake the work of community organizing, you are helping others find the heroes within themselves.

To determine if you are a good fit for being a community organizer, consider:

- How much time — and for how long — can you commit to organizing?
- Do you have the core abilities and skills to:
  - Communicate the climate emergency message?
  - Recruit and manage others?
  - Plan events?
  - Use databases and social media?
  - Speak in public comfortably and effectively?

- If you lack some of these skills, can you build a team that does?
- Who do you know and respect the most? Who would you most want to join you in the fight for humanity and all life?
  - Which business and religious leaders, philanthropists, influencers, tastemakers, celebrities, authors, and thought-leaders can you recruit into the movement?
  - Do you know any elected leaders and their staffers? Can you identify them and bring them into the movement?

Start a Campaign or Organization. If you have 25 hours or more a week and a big idea for the climate emergency movement, you might consider starting a new campaign or organization. I certainly understand the impulse — I made this choice five years ago because I could not find an existing organization with a firm commitment to telling the truth and demanding a solution that could actually protect humanity and the natural world. But take it from me, this path poses many challenges. It is much easier to work within an existing structure, with an existing team. When you start a new organization from the ground up, all of the problems and least-popular jobs fall to you, from handling internal conflict and facing external criticism to fundraising and paperwork. I certainly don't want to dissuade you from the challenge of executing your vision to stop our climate emergency, but first consider the possibility that your project may already fit within another organization or that you could convince another organization to take on your project.

However, if you have a unique vision for how your organization can save the world, you are 100 percent dedicated to making it happen, and you have 25 hours a week (or more) to dedicate, then you should do it. All of the organizations I've discussed — TCM, Extinction Rebellion, Sunrise Movement, the Future Coalition, the Justice Democrats — are less than five years old. One person, or a few people, started each of these organizations with a vision and belief that they could realize it.

Independent projects can also be worthwhile. End Climate Silence, Genevieve Guenther's project that uses Twitter to call out reporters, editors, and publications for failing to mention the impact of the climate emergency in highly relevant stories, is less than two years old and has already had a major impact. The concept is simple, but it gets results. Some of Guenther's call-outs of media negligence get retweeted thousands of times, and often the stories are subsequently edited to include climate impacts. Greta Thunberg's school climate strike also started out as an independent project, before growing into the global school strike movement.

Maybe the project you want to launch is a campaign for elected office. Running for office on a climate emergency platform or supporting a campaign that runs on one can be a highly effective way to spread the climate emergency message and, hopefully, bring the argument for climate mobilization directly into government. Many resources are available to guide your pursuit of local or state offices, including those at Run for Something.

Perhaps your organization or project is the next big breakthrough. How will you know? How will you be able to decide whether you are on the right track; whether your efforts are working? *The Lean Startup* by Eric Ries was invaluable to me in answering those questions. Ries proposes an ongoing process of identifying, testing, and proving your assumptions as your organization grows. This process can help prevent you from needlessly spending money and time on projects that don't get results — a webpage, for example, that nobody reads.

If you think that starting something new might be the heroic path for you, answer the following questions:

- What is your idea for saving the world?
- Can you pursue your project under the umbrella of another organization?

- Do you have evidence that your idea is a good one? Are you sure it appeals to other people?
- How can you test your idea to determine its effectiveness?
- Do you have friends and family who will give their honest feedback about your idea, and who can tell you about your strengths and weaknesses in leading new ventures?

Volunteer Special Skills. Most people join the climate emergency movement in some kind of community organizing role, as discussed previously. But there are also plenty of jobs for specialists. For example, TCM has a volunteer research team, social media team, graphic designer, fundraisers, copy editors, and more. The time commitment will vary based on the skill in question and the level of coordination necessary.

Below is a list of specialized tasks that climate emergency organizations almost always need accomplished and would very often be happy to have a volunteer perform. (Fundraising is so important and necessary that it deserves its own section, which follows.)

- Event planning
- Project management
- General administrative support
- Publicity and press support
- Graphic design; web and user-interface design
- Database design and management
- Tech support
- Social media
- Bookkeeping and accounting
- Legal advising
- Grant-writing
- Copyediting

- Cooking/catering
- Video and audio production
- Research
- Human resources and recruiting support
- Childcare

If you have an interest in or specific skills that apply to these tasks, reach out and offer your time to your climate emergency organization of choice.

Another approach to applying your skills is by supporting *organizers* rather than directly supporting an organization. This requires less coordination and management. If you are a professional who offers clients services, consider making a portion of your work pro bono or available at a reduced fee for people in the climate emergency movement. Even a few hours per week can be extremely helpful to organizers. For example, tax preparers can offer guidance to individual organizers as well as organizations. Lawyers can represent activists who have been arrested. Real-estate brokers can help organizers find situations for affordable co-housing. Or, you might be a professional who offers services that can help organizers avoid burnout. Massage therapists can offer their services, as they've done for organizers at Extinction Rebellion headquarters. Mental health professionals can offer individual or group therapy to organizers. Whatever your area of expertise, consider how organizers could benefit from it.

To identify the skills you can bring to volunteer work, answer the following questions:

- What skills do you have that organizers or organizations could benefit from?
- How many hours per week can you volunteer, and how flexibly?
- How will you effectively communicate to organizations or organizers that your skills are available?

Fundraise. Fundraising is the volunteer task that is arguably the highest leverage, but is massively unpopular because it's emotionally challenging and can be awkward. Want to make a big impact in the movement? Fundraise.

Some people are under the illusion that social movement organizations shouldn't need money. They think these organizations should operate exclusively with volunteers and miracles. This is a beautiful concept — and it sometimes works for a limited period of time, as it did during the Occupy Wall Street. However, while Occupy riveted the public's attention, it could not turn its groundswell into concrete policy change. For a movement to do that, it needs to sustain itself over months and years, and it typically needs full-time staff, including coordinators, managers, and volunteers. Social movement organizations need people who are available for meetings during the workweek. They need people who will, consistently and diligently, do the things no one else wants to do, like handling the legalities, dealing with difficult people and situations, and keeping the organization financially solvent. The staff doesn't necessarily need to be well-paid — Extinction Rebellion covers only the barest of its core team's living expenses — but they do need something. Then there are other expenses — office space, travel, and printing, to name a few. Simply put, organizations need money!

For organizations that are considered "radical" and are outside the comfort zone of traditional philanthropic funders, securing funding can be extremely challenging.[154] This is why almost any organization will say, "Yes, please!" to someone offering to work as a fundraising volunteer, especially if that person can fundraise from within their own networks. Fundraising can take many forms. It might mean inviting friends and family over for a discussion about the climate crisis and, at the discussion's end, asking for donations to your favored organization. It might mean asking for money through your social media presence. Facebook, for example, has a platform where you can fundraise for nonprofit organizations. It might mean that you "pass the hat" at work, or that you email 10 friends to tell them you are making a donation and asking them to match it. Climate emergency organizations might also ask fundraising volunteers to call past donors or people who are supporters, but not donors, and ask them for money; to do background research on donors; or to help prepare grant applications.

If you think you might be able to fundraise effectively, consider reading Kim Klein's classic fundraising text, *Fundraising for Social Change*.[155] Klein talks about getting over the fear of asking for money by recognizing that the cause is more important than the fear. Indeed, fundraising for the climate emergency movement is heroic — a challenging but crucial part of our fight for civilization and the natural world. To assess your fundraising preparedness, answer the following questions:

- Have you ever asked someone to donate money to a cause? How comfortable or uncomfortable were you?
- What about fundraising is appealing or repellent?
- To solicit donations, are you willing to overcome any social awkwardness and fears about asking?
- How wealthy and philanthropic is your professional and social network? Do they regularly donate to social causes?

Klein encourages fundraisers to view soliciting donors as offering them an opportunity to be a part of something meaningful. In this case, giving to the movement isn't charity, it's enlightened self-interest. Despite the hopes of space-exploring, luxury-bunker-building billionaires, there is no place to hide from the climate crisis. Our only hope, our only option, is to transform the world. How could money be put to better use?

On that note — please donate! Our culture tells us that building wealth is building security for ourselves and our families.

Money is safety. Money is freedom. A friend told me that his only moral obligation in the face of the crisis was to "build wealth for my family." That view is utterly nonsensical in the age of ecological breakdown. We're told to save "for the future," but the ecological crisis is destroying that future.

Individuals, institutions, and governments all need to deploy the financial resources we have to protect humanity and restore a safe climate, now. This is what emergency mode means, not hoarding money for an imagined, denial-based future, but rather mobilizing our resources to stop the emergency and restore safety. We cannot let civilization collapse while our money sits in banks or the stock market. Philanthropies, institutions, and individuals like you and me must give aggressively to organizations and candidates who support emergency climate mobilization. When the mobilization arrives, the government should spend without limit to save as much life as possible. Deploying your financial resources now for the climate emergency movement is one of the most impactful ways you can possibly use your money.

Supporting the movement financially is not only a moral duty, it is also the best way to achieve individual and familial security. True security cannot coexist with the climate emergency and the sixth mass extinction of species. True security can be achieved only through the restoration of a safe climate and the natural world. Therefore, not only do I suggest that you give money to the climate emergency movement, I also suggest that you hold all organizations and politicians to a climate emergency standard for giving. This means that if you are a contributor to the Natural Resources Defense Council, the Environmental Defense Fund, or even an "unrelated" organization like an art museum or a university, you tell fundraisers and others that you will support them only if they declare a climate emergency and are actively working to protect humanity and the natural world from catastrophe. Hold this line for politicians, too. If they aren't making the climate emergency their top priority and advocating

WWII-scale climate mobilization that eliminates emissions in 10 years or less, stop supporting them — and tell them why.

To better prepare to contribute your money to the climate emergency movement, ask yourself if you can imagine making a financial commitment that feels commensurate with the scale and urgency of the climate crisis itself. If the answer is yes — and it must be — then consider these quesitons:

- How much money do you have saved?
- What is your net worth?
- What is your annual income?
- How much do you spend a year, and what are your biggest expenses?
- What one-time donation could you make that would push your capacity for giving?
- What maximum monthly contribution could you commit to for one year?
- What lifestyle sacrifices could you make to donate more money to climate change mobilization?
- What other tangible or intangible resources can you commit (a meeting space, a vehicle, etc.)?
- What climate emergency organizations are you most impressed by? How can your money help their efforts?

Get Started. We must respond to the climate emergency not only with our heads and hearts; we have a moral obligation to respond with dedicated and fierce action. Reflect on where you fit in, and consider your "highest and best use," but don't get stuck in analysis paralysis.

Determining your place in the movement is an ongoing process. Have a conversation with someone about the climate emergency today and every day. Make a financial contribution today, and then start thinking about your next contribution. Attend a meeting this week, and grow your role from there. If you still aren't sure where to begin, visit TCM's website (www.TheClimateMobilization.org) to fill out an online questionnaire that makes recommendations about where you might fit into the movement (www.TheClimateMobilization.org/Transformation). Above all, remember that your actions are contagious and can have a huge impact on others. By breaking the silence and entering emergency mode, you will lead your friends, family members, coworkers, and neighbors to join you.

Questions for Reflection and Discussion

- How many hours a week will you commit to the climate emergency movement?
- What are the biggest barriers to dedicating even more time?
- What potential roles seem the most appealing to you? Which roles do you want to avoid?
- What do you think your highest and best use might be to the movement?
- Would you consider sharing this book with others? For whom might it be helpful?

CONCLUSION:
Live as a Climate Warrior

THIS IS MEANT TO BE A PRACTICAL BOOK. It is meant to support your becoming a more effective change-agent in initiating the "great awakening" to save humanity and life on Earth. Thank you for your courage in confronting the pain of the climate emergency and for seriously considering how to maximize your contribution.

Dedicating yourself to canceling the apocalypse and protecting all life is the most challenging thing you will ever do. When you set your ambitions on saving the world, there is always more to do. Your career and finances may suffer. Your relationships may be strained. You will certainly experience setbacks and frustrations. Do it anyway. Even though you don't have every skill, even though you have faults, you are good enough to do this work. By dedicating your life to a safe climate, you contribute to protecting all life. You matter. Be proud of your courage and contributions.

Neoliberalism has told us that we are self-interested consumers who are cynical and passive about politics. The crisis we are facing demands a different kind of person — someone who takes *personal responsibility* for solving the global climate and ecological crises, and is dedicated to spreading climate truth and initiating full-scale climate mobilization. We need people to understand the fundamental interconnectedness of all life, and are ready to fight for what they love. We need people who are unwilling to lie to themselves, who are ready to be transformed by the truth, and then lead others in that transformation.

In 1943, while the home-front mobilization was underway, American psychologist Abraham Maslow developed his famous hierarchy of needs.[156]

Rather than study mental illness, Maslow investigated how humans flourish. He believed that people could achieve their highest potential only by first addressing their basic needs. Before we can be free to meet our needs for safety, esteem, and love and belonging, we must first meet our needs for food, clothing, and shelter. Even if we satisfy all of our basic needs, we can expect a new discontent and restlessness to develop unless we pursue our potential. If they are to be ultimately happy, a musician must make music; a carpenter must build; a healer must heal. What a person can be, they must be. Maslow refers to this need as self-actualization.

Self-actualization may sound self-centered. It is anything but. It is through meeting our basic needs and tapping into our truest selves and potential that we are able to best serve something greater than ourselves. Maslow studied self-actualized people and wrote about their characteristics. Far from characterizing them as self-absorbed, he described them as "problem-centered." Self-actualized people have a mission in life, some task to fulfill, some problem outside of themselves that enlists much of their

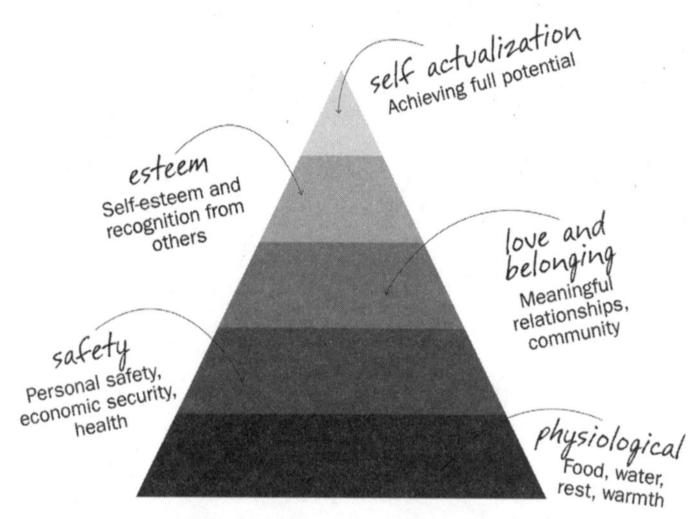

Maslow's Hierarchy of Needs, CREDIT: JEFFREY SAKS

energies. These aren't necessarily tasks that they would prefer or choose for themselves. They may be tasks that they feel are their responsibilities, duties, or obligations.[157]

In this time of acute ecological crisis, being creatively, passionately, and humbly engaged in the climate emergency movement is the most important, and perhaps the *only* way to be truly self-actualized. We didn't choose this emergency; we would probably rather be doing something else. But here we are, in personal and collective danger. The climate emergency chose us, and we must devote our vast potential to reversing it.

It is an honor and a privilege to walk the path of the climate warrior, but it's a privilege that is unavailable to many. Across the country and globe, many people already lack food, security, and shelter. Wealth inequality, combined with the climate emergency, makes life extremely difficult for many. As the Earth continues to warm, people who now claim only precarious security will likely face even bigger challenges. People who are in danger of being evicted or going hungry cannot be expected to prioritize global cataclysm ahead of their daily existential risk. Others, by virtue of their location, religion, or race, will never be assured of personal safety. Still others may lack a tightly woven familial or social safety net to support them as they work to cope with the pain of the climate emergency and the challenges of joining the movement. Simply put, few people have the freedom to pursue self-actualization through becoming climate warriors. If you can, you *must.*

When you reach for self-actualization, you will be awed by the immensity and beauty of all life. You will be grateful and proud to be a warrior in its service. You will take your rightful place beside your teammates and allies. With every step you take, you fulfill your destiny as a protector of humanity and all life.

Transformation is hard work — painful, almost impossibly hard work — but nature promises us that it's worthwhile. When we undertake the work of transformation, we don't merely become "better," we become something entirely new. Environmentalists sometimes draw inspiration from one of the most dramatic transformational processes in nature — a caterpillar becoming a butterfly. As environmentalist Kim Polman writes in *Imaginal Cells: Visions of Transformation,*

> After a period of ravenous consumption, the caterpillar forms a chrysalis from which it will dissolve itself into an organic stew, where dormant "imaginal cells" hold the vision of the new structure. At first, these imaginal cells operate independently, as single-cell organisms, and are attacked by the caterpillar's immune system, which views them as a threat. But soon, these new cells regroup, multiply, and connect with each other. They then form clusters and begin resonating at the same frequency. Finally, they reach a tipping point and consolidate to become a new multicellular organism, the beautiful butterfly.[158]

The caterpillar consumes voraciously, but its imaginal cells lead to its transformation into a butterfly — a pollinator and a powerful contributor to the health of the entire ecosystem. What the caterpillar takes, the butterfly gives back. Humanity, especially those of us in the United States and other affluent countries, has consumed voraciously. But humanity is capable of so much more. It's time to transform and give back. It's time to engage in the emergency mobilization that will protect humanity and create a society of givers, protectors, and healers. It's time to shift from a society based on destruction to one based on regeneration.

Now is our time to become humanity's imaginal cells and lead the transformation. We have been isolated and dormant for too long, but now we are organizing, growing, and building a movement.

We know we will face the parts of the old system that are afraid, the parts that view change as a threat. But we will withstand attacks because we know that our work to lead society to the tipping points is necessary for transformation.

It is not a given that we will successfully transform, that this movement will win with enough time to avert civilization's collapse, but it is our only hope. We must join together to do everything we can to initiate maximum intensity mobilization as quickly as possible. We must turn our pain into action and take personal responsibility for protecting humanity and the living world. We must become heroes. Onward!

ADDITIONAL RESOURCES:
Continue Your Journey

Share this Book, Free!

DON'T ALLOW THIS BOOK TO SIT IDLY on your shelf — lend or gift a copy to a friend *as soon as you possibly can.* On our website, you can **send the first chapter of this book to a friend for free at www.TheClimateMobilization.org/ShareTheBook.** We also offer a book club program, and group/bulk-ordering discounts.

What's Next?

It's been an honor to take this journey with you. Thank you for your courage, vulnerability, and commitment. We have shared a great deal together, and I would welcome the opportunity to continue our conversation and our work.

Please join me in the ***Climate Truth Transformation Hub*** to:

- Share your reaction to this book with me. I'd love to hear from you!
- Use our self-assessment tools to identify the highest-impact climate actions you can take right now based on your skills and resources.
- Connect with other climate warriors for ongoing motivation and support to process difficult emotions and live more fully in climate truth.
- Access additional content, including new articles, video trainings, step-by-step guides to help you break the silence, author Q&A sessions, and more.

All online resources are free. Visit: www.TheClimateMobilization.org/Transformation

ENDNOTES

Introduction

1. Jared Keller, "The U.S. Suicide Rate Is at Its Highest in Half a Century," P/S, December 4, 2018, https://psmag.com/news/the-suicide-rate-is-at-its-highest-in-a-half-century (accessed July 23, 2019).
2. Thomas Moore and Donald Mattison, "Adult Utilization of Psychiatric Drugs and Differences by Sex, Age, and Race," *JAMA Internal Medicine* 177 (February 2017): 274–275, https://jamanetwork.com/journals/jamainternalmedicine/fullarticle/2592697 (accessed July 23, 2019).
3. National Safety Council: Injury Facts, "Preventable Deaths," https://injuryfacts.nsc.org/all-injuries/preventable-death-overview/odds-of-dying/data-details/ (accessed July 23, 2019).
4. Nielsen, "The Total Audience Report: Q1 2016," https://www.nielsen.com/us/en/insights/report/2016/the-total-audience-report-q1-2016/ (accessed July 23, 2019).
5. Genevieve Guenther, "Who Is the 'We' in 'We Are Causing Climate Change'?" *Grist,* October 13, 2018, https://grist.org/article/who-is-the-we-in-we-are-causing-climate-change/ (accessed July 23, 2019).
6. Erich Fromm, "Love of Death and Love of Life," *The Heart of Man: Its Genius for Good and Evil,* ed. Ruth Nanda Anshen (New York: Harper, 1964).
7. Erich Fromm, "Credo," *Erich Fromm Online,* https://fromm-online.org/en/das-leben-erich-fromms/fromms-credo-eines-humanisten/ (accessed July 24, 2019).
8. Ibid.
9. Naomi Klein, *This Changes Everything: Capitalism vs. The Climate* (New York: Simon & Schuster, 2014).
10. D.J. Peterson, *Troubled Lands: The Legacy of Soviet Environmental Destruction.* (Boulder: Westview Press, 1993).
11. Ibid., 36; Global Carbon Project, "CO_2 Emissions," Global Carbon Atlas, http://www.globalcarbonatlas.org/en/CO2-emissions (accessed November 12, 2019).

12 Norwegian Ministry of Trade, Industry and Fisheries, *The State Ownership Report 2015* (Norway: Norwegian Ministry of Trade, Industry and Fisheries, 2015), https://www.equinor.com/content/dam/statoil/documents/the-state-ownership-report-2015.pdf (accessed August 18, 2019).

13 Edward O. Wilson, *Half-Earth: Our Planet's Fight for Life* (New York: Liveright, 2016).

14 Pope Francis, *Encyclical Letter Laudato Sí' of the Holy Father Frances On the Care of Our Common Home*, May 24, 2015, http://w2.vatican.va/content/francesco/en/encyclicals/documents/papa-francesco_20150524_enciclica-laudato-si.html (accessed October 23, 2019).

Step One: Face Climate Truth

15 David Wallace Wells, *The Uninhabitable Earth: Life After Warming* (New York: Penguin Random House, 2019).

16 Megan Darby, "Meet the Woman Who First Identified the Greenhouse Effect," *Climate Change News*, February 9, 2016, https://www.climatechangenews.com/2016/09/02/the-woman-who-identified-the-greenhouse-effect-years-before-tyndall/(accessedJuly24, 2019).

17 Marit-Solveig Seidenkrantz, "80 Years Since the First Calculations Showed that the Earth Was Warming Due to Rising Greenhouse Gas Emissions," *Science X*, June 5, 2018, https://phys.org/news/2018-06-years-earth-due-greenhouse-gas.html (accessed July 24, 2019).

18 Neela Banerjee, Lisa Song, and David Hasemyer, "Exxon's Own Research Confirmed Fossil Fuels' Role in Global Warming Decades Ago," *Inside Climate News*, September 16, 2015, https://insideclimatenews.org/news/15092015/Exxons-own-research-confirmed-fossil-fuels-role-in-global-warming (accessed July 24, 2019).

19 Naomi Oreskes and Erik Conway, *Merchants of Doubt: How a Handful of Scientists Obscured the Truth on Issues from Tobacco Smoke to Global Warming* (New York: Bloomsbury, 2010).

20 Katie Worth, "Climate Change Skeptic Group Seeks to Influence 200,000 Teachers," *Frontline* PBS.org, https://www.pbs.org/wgbh/frontline/article/climate-change-skeptic-group-seeks-to-influence-200000-teachers/ (accessed October 23, 2019).

21 Robert J. Brulle, "Institutionalizing Delay: Foundation Funding and the *Creation of U.S. Climate Change Counter-Movement Organizations,"*

Climatic Change: An Interdisciplinary, International J*ournal Devoted to the Description, Causes and Implications of Climatic Change,* 122,4 (2014): 681–694.

22 Robert J. Brulle, "The Climate Lobby: A Sectoral Analysis of Lobbying Spending on Climate Change in the USA, 2000 to 2016," *Climatic Change : An Interdisciplinary, International Journal Devoted to the Description, Causes and Implications of Climatic Change,* 149, 3–4 (2018): 289–303.

23 Jonathan Chait. "Why Are the Republicans the Only Climate-Science Denying Party in the World?" *New York Magazine,* September 27, 2015, http://nymag.com/intelligencer/2015/09/whys-gop-only-science-denying-party-on-earth.html?gtm=top (accessed July 25, 2019).

24 "Leaders: A Greener Bush," *The Economist,* February 15, 2003, https://www.economist.com/leaders/2003/02/13/a-greener-bush (accessed August 18, 2019).

25 Valerie Richardson, "Obama Takes Credit for U.S. Oil-and-Gas Boom: 'That Was Me, People,'" *The Washington Times,* November 28, 2018, https://www.washingtontimes.com/news/2018/nov/28/obama-takes-credit-us-oil-and-gas-boom-was-me-peop/ (accessed August 18, 2019).

26 Glenn Scherer, "Climate Science Predictions Prove Too Conservative," *Scientific American,* December 6, 2012, https://www.scientificamerican.com/article/climate-science-predictions-prove-too-conservative/ (accessed on July 26, 2019).

27 David Spratt and Ian Dunlop, *What Lies Beneath: The Understatement of Existential Climate Risk* (Melbourne, Australia: Breakthrough, 2018), https://docs.wixstatic.com/ugd/148cb0_a1406e0143ac4c469196d3003bc1e687.pdf.

28 Michael Mann, Susan Hassol, and Tom Tole, "Doomsday Scenarios Are as Harmful as Climate Change Denial," *The Washington Post,* July 12, 2017, https://www.washingtonpost.com/opinions/doomsday-scenarios-are-as-harmful-as-climate-change-denial/2017/07/12/880ed002-6714-11e7-a1d7-9a32c91c6f40_story.html?utm_term=.c6bff330726d (accessed July 26, 2019).

29 Anand Giridharandas, *Winners Take All: The Elite Charade of Changing the World* (New York: Knopf, 2018).

30 Debika Shome and Sabine Marx, *The Psychology of Climate Change Communication: A Guide for Scientists, Journalists, Educators, Political Aides, and the Interested Public.* (New York: Center for Research on Environmental Decisions, 2009).

31 Shome and Marx, *The Psychology of Climate Change Communication.*

32 Daniel Chapman, Brian Lickel, and Ezra Markowitz. "Reassessing Emotion in Climate Change Communication," Nature 7 (2017): 850–852.

33 "A Field Guide to the Climate Movement," *Inside Climate News,* April 8, 2015, https://insideclimatenews.org/content/field-guide-us-environmental-movement (accessed August 10, 2019).

34 Michael Tobis and Stephen Ban, "OK, Getting Serious Again," *Only in It for the Gold* (blog), Planet 3, January 14, 2010, http://init.planet3.org/2010/01/ok-getting-serious-again.html (accessed July 29, 2019).

35 DARA and the *Climate Vulnerable Forum. Climate Vulnerability Monitor: A Guide to the Cold Calculus of a Hot Planet,* 2nd ed. (Spain: DARA, 2012), https://daraint.org/wp-content/uploads/2012/09/CVM2ndEd-FrontMatter.pdf (accessed July 26, 2019).

36 David Spratt and Ian Dunlop, *Existential Climate Risk: A Scenario Approach* (Melbourne, Australia: Breakthrough, 2019), https://docs.wixstatic.com/ugd/148cb0_a1406e0143ac4c469196d3003bc1e687.pdf.

37 IPCC, *Global Warming of 1.5°C: An IPCC Special Report on the Impacts of Global Warming of 1.5°C Above Pre-Industrial Levels and Related Global Greenhouse Gas Emission Pathways,* ed. V. Masson-Delmotte, P. Zhai, H.O. Pörtner, D. Roberts, J. Skea, P.R. Shukla, A. Pirani (In Press, 2018), https://www.ipcc.ch/site/assets/uploads/sites/2/2019/05/SR15_SPM_version_report_LR.pdf (accessed July 26, 2019).

38 "Planetary Boundaries Research," *Stockholm Resilience Centre,* https://www.stockholmresilience.org/research/planetary-boundaries.html (accessed July 29, 2019).

39 "The Extinction Crisis," *Center for Biological Diversity,* https://www.biologicaldiversity.org/programs/biodiversity/elements_of_biodiversity/extinction_crisis/ (accessed August 10, 2019).

40 M. Grooten and R.E.A. Almond, eds., *Living Planet Report 2018: Aiming Higher* (Switzerland: WWF, 2018).

41 Caspar Hallman et al., "More than 75 Percent Decline Over 27 Years in Total Flying Insect Biomass in Protected Areas," *PloS ONE* 12,10 (2017): e0185809, https://doi.org/10.1371/journal.pone.0185809.

42 Cheryl Schultz, Leone Brown, Emma Pelton, and Elizabeth Crone, "Citizen Science Monitoring Demonstrates Dramatic Declines of Monarch Butterflies in Western North America," *Biological Conservation*

214 (2017): 343–346, https://doi.org/10.1016/j.biocon.2017.08.019.
43 Paul Gilding, *The Great Disruption* (London: Bloomsbury Press, 2011).
44 D. Lin, L. Hanscom, A. Murthy, et al., "Ecological Footprint Accounting for Countries: Updates and Results of the National Footprint Accounts, 2012–2018," *Resources* 2018, 7, 58.
45 Michael Borucke, David Moore, Gemma Cranston, et al., "Accounting for demand and supply of the biosphere's regenerative capacity: The National Footprint Accounts' underlying methodology and framework," *Ecological Indicators*, 24, January 2013, 518–533.
46 Lester Brown, "Could Food Shortages Bring Down Civilization?" *Scientific American* (May 2009): 50–57, http://www.earth-policy.org/images/uploads/press_room/SciAm-final.pdf (accessed July 29, 2019).
47 Peter H. Gleick, "Water, Drought, Climate Change, and Conflict in Syria," *Weather, Climate, and Society* 6, 3 (2014): 331–34.
48 European Commission, "Worrying Effects of Accelerating Climate Change on the Mediterranean Basin," *EU Science Hub,* October 22, 2019, https://ec.europa.eu/jrc/en/science-update/worrying-effects-accelerating-climate-change-mediterranean-basin (accessed August 1, 2019).
49 United Nations Security Council, "Statement by the President of the Security Council," United Nations, January 30, 2018, https://www.un.org/en/ga/search/view_doc.asp?symbol=S/PRST/2018/3 (accessed August 1, 2019).
50 Bill McKibben, *Eaarth: Making a Life on a Tough New Planet* (New York: Times Books, 2010):
51 Ibid., 2
52 Ibid., 5.
53 Robert B. Cialdini, *Influence: Science and Practice,* 5th ed (Harlow, Essex: Pearson, 2008).
54 John M. Darley and Bibb Latané, "Bystander Intervention in Emergencies: Diffusion of Responsibility," *Journal of Personality and Social Psychology* 8, 4 (1968): 377–83.
55 Bibb Latané and John M. Darley, *Group Inhibition of Bystander Intervention in Emergencies* (Emmitsburg, MD: National Emergency Training Center, 1968).
56 D. Nilsson and A. Johansson, "Social Influence during the Initial Phase of a Fire Evacuation — Analysis of Evacuation Experiments in a Cinema Theatre," *Fire Safety Journal* 44 (2008): 71–79.

57 Doris Goodwin, "The Way We Won: America's Economic Breakthrough During WWII," *American Prospect* (Fall 1992), https://prospect.org/article/way-we-won-americas-economic-breakthrough-during-world-war-ii (accessed July 29, 2019).

58 "Table 3.1," *Historical Tables: Budget of the United States Government Fiscal Year 2011,* White House Office of Management and Budget, https://www.whitehouse.gov/omb/historical-tables/ (accessed August 18, 2019).

59 Price V. Fishback and Joseph Cullen, *Did Big Government's Largesse Help the Locals? The Implications of WWII Spending for Local Economic Activity,* 1939–1958 (Cambridge: National Bureau of Economic Research, 2006).

60 Douglas Keay, "Interview for *Women's Own,*" Margaret Thatcher Foundation, October 16, 1984, https://www.margaretthatcher.org/document/105577.

61 John Duberstein, *A Velvet Revolution: Václav Havel and the Fall of Communism* (Greensboro, North Carolina: Morgan Reynolds, 2006).

62 Václav Havel, *The Power of the Powerless: Citizens Against the State in Central Eastern Europe,* trans. John Keane (Abingdon: Routledge, 1985).

63 United Planet Faith & Science Initiative, "Greta Thunberg: Our House Is on Fire. 2019 World Economic Forum (WEF) in Davos," YouTube video, 6:03. January 25, 2019. https://www.youtube.com/watch?v=zrF1THd4bUM (accessed August 18, 2019).

Step Two: Welcome Fear, Grief, and Other Painful Feelings

64 Kerry Kelly Novick and Jack Novick, *Emotional Muscle: Strong Parents, Strong Children* (Xlibris: Author, 2010).

65 Kristen Neff, *Self-Compassion: The Proven Power of Being Kind to Yourself* (New York: HarperCollins, 2011).

66 Benjamin Chapman, Kevin Fiscella, Ichiro Kawachi, et al., "Emotional Suppression and Mortality Risk over a 12-Year Follow-Up," *Journal of Psychosomatic Research,* 75, 4 (2013): 381–385, https://www.ncbi.nlm.nih.gov/pmc/articles/PMC3939772/.

67 Jean Carlomusto, dir., *Larry Kramer: In Love & Anger* (2015).

68 Harriet Sherwood and Angela Guiffrida, "Pope Reveals He Had Weekly Psychoanalysis Sessions at Age 42," *The Guardian,* September 1, 2017, https://www.theguardian.com/world/2017/sep/01/pope-francis-psychoanalysis (accessed July 31, 2019).

69 "RAIN: Recognize, Allow, Investigate, Nurture," Tara Brach (website), https://www.tarabrach.com/rain/ (accessed August 19, 2019).

70 J. Rottenberg, F.H. Wilhelm, J.J. Gross, and I.H. Gotlib, "Vagal rebound during resolution of tearful crying among depressed and nondepressed individuals," *Psychophysiology* 40:1–6.

71 William H. Frey and Muriel Langseth. *Crying: The Mystery of Tears* (Minneapolis, MN: Winston Press, 1985).

72 Asmir Gračanin, Lauren Bylsma, and Ad J.J.M. Vingerhoets, "Is Crying a Self-Soothing Behavior?" *Frontiers in Psychology* 5 (May 2014), http://d-scholarship.pitt.edu/24767/1/fpsyg-05-00502.pdf.

73 N. van Leeuwen, E.R. Bossema, H. van Middendorp, et al., "Dealing with Emotions When the Ability to Cry is Hampered: Emotion Processing and Regulation in Patients with Primary Sjögren's Syndrome," *Clinical and Experimental Rheumatology*, Jul-Aug; 30 (4):492–8, http://hdl.handle.net/2066/110788.

74 A.J.J.M. Vingerhoets, N. van de Ven, and Y. van der Velden. "The Social Impact of Emotional Tears," *Motivation and Emotion* 40 (3): 455–463, doi: 10.1007/s11031-016-9543-0.

75 Jaak Panksepp, "Affective Neuroscience of the Emotional BrainMind: Evolutionary Perspectives and Implications for Understanding Depression," *Dialogues in Clinical Neuroscience* 12, 4 (Dec. 2010): 533–45.

76 Thierry Steimer, "The Biology of Fear- and Anxiety-Related Behaviors," *Dialogues in Clinical Neuroscience* 4, 3 (Sept. 2002): 231–49.

77 Joanna Macy and Chris Johnstone, *Active Hope: How to Face the Mess We're in Without Going Crazy* (California: New World Library, 2012): 71.

78 Ibid.

79 DARA and the Climate Vulnerable Forum, *Climate Vulnerability Monitor, 2nd Edition: A Guide to the Cold Calculus of a Hot Planet* (Madrid: Fundación DARA Internacional, 2012).

80 WWF, *Living Planet Report 2016: Risk and resilience in a new era* (WWF International, Gland, Switzerland, 2016), https://c402277.ssl.cf1.rackcdn.com/publications/964/files/original/lpr_living_planet_report_2016.pdf?1477582118&_ga=1.148678772.2122160181.1464121326.

81 Mary Annaïse Heglar, "The Big Lie We're Told about Climate Change Is that It's Our Fault," *Vox*, November 27, 2018, https://www.vox.com/first-person/2018/10/11/17963772/climate-change-global-warming-natural-disasters (accessed July 31, 2019).

82 Carolyn Baker and Guy R. McPherson, *Extinction Dialogues: How to Live with Death in Mind* (Next Revelation Press, 2014).

83 Carolyn Baker, *Collapsing Consciously: Transformative Truths for Turbulent Times* (Berkeley, CA: North Atlantic Books, 2013).
84 Macy and Johnstone, *Active Hope*: 75.
85 Ibid.
86 Ibid.
87 Heglar, "The Big Lie We're Told about Climate Change.
88 Wikipedia contributors, "Theodore Parker," *Wikipedia, The Free Encyclopedia*, https://en.wikipedia.org/w/index.php?title=Theodore_Parker&oldid=918020874 (accessed October 31, 2019).
89 Wells, *The Uninhabitable Earth*.

Step Three: Reimagine Your Life Story

90 Margaret Salamon, Dissertation: The Trauma of a Romantic Partner's Psychotic Episode: An Emerging Clinical Picture. Gordon Derner Institute of Advanced Psychological Studies, Adelphi University. Margaret E. Klein September, 2014.
91 Edward O. Wilson, *The Social Conquest of Earth* (New York: Liveright, 2012): 241.
92 Martin Luther King, Jr., "Letter from a Birmingham Jail", August 1963, http://web.cn.edu/kwheeler/documents/letter_birmingham_jail.pdf (accessed August 1, 2019).

Step Four: Understand and Enter Emergency Mode

93 Doris Kearns Goodwin, *No Ordinary Time: Franklin & Eleanor Roosevelt: The Home Front in WW II* (New York: Simon & Schuster, 1994); David Kaiser, *No End Save Victory: How FDR Led the Nation into War* (Philadelphia: Basic Books, 2014); Arthur Herman, *Freedom's Forge: How American Business Produced Victory in World War II* (New York: Random House, 2013).
94 M.M. Eboch, *Native American Code Talkers* (Minneapolis, MN: ABDO, 2015).
95 Goodwin, *No Ordinary Time*; Kaiser, *No End Save Victory*; Herman, *Freedom's Forge*, 332.
96 Thomas Bassett, "Reaping on the Margins: A Century of Community Gardening in America," *Landscape* 25, 2 (1981): 1–8.
97 "Table 3.1," *Historical Tables: Budget of the United States Government Fiscal Year 2011*, White House Office of Management and Budget, https://www.whitehouse.gov/omb/historical-tables/(accessed August 18, 2019).

98 Lester Brown, *Plan B 4.0: Mobilizing to Save Civilization* (New York: WW Norton, 2009): 260.
99 Maury Klein, *A Call to Arms: Mobilizing America for World War II* (New York: Bloomsbury Publishing, 2015).
100 Hugh Rockoff, *Keep on Scrapping: The Salvage Drives of World War II* (Cambridge, Mass: National Bureau of Economic Research, 2007), http://papers.nber.org/papers/w13418.
101 National Science Foundation, *The National Science Foundation: A Brief History* (Washington, D.C.: ERIC Clearinghouse, 1988).
102 Andrew Edmund Kersten, *Race, Jobs, and the War: The FEPC in the Midwest, 1941–46* (Urbana: University of Illinois Press, 2007).
103 Ronald T. Takaki, *Double Victory: A Multicultural History of America in World War II* (Boston: Little, Brown and Company, 2001).
104 Goodwin, *No Ordinary Time*, 555.
105 Ibid, 769.
106 Ibid, 628.
107 Lydia Saad, "Gallup Vault: A Country Unified After Pearl Harbor," Gallup, December 5, 2016, https://news.gallup.com/vault/199049/gallup-vault-country-unified-pearl-harbor.aspx (accessed August 18, 2019).
108 Ibid., 233.
109 Kaiser, *No End Save Victory*.
110 Ibid.
111 Goodwin, *No Ordinary Time*, 356–59.
112 David Spratt and Philip Sutton, *Climate Code Red: The Case for Emergency Action* (Carlton North, Victoria, Australia: Scribe Publications, 2009).
113 Goodwin, "The Way We Won."
114 Klein, *A Call to Arms*; Goodwin, "The Way We Won."
115 Joe Stiglitz, "The Climate Crisis Is Our Third World War: It Needs a Bold Response," *The Guardian*, June 4, 2019, https://www.theguardian.com/commentisfree/2019/jun/04/climate-change-world-war-iii-green-new-deal (accessed August 18, 2019).
116 Thomas Friedman, *Hot, Flat, and Crowded: Why We Need a Green Revolution — and How It Can Renew America* (New York: Farrar, Straus and Giroux, 2008).
117 Bill McKibben, "A World At War," *The New Republic*, August 15, 2016, https://newrepublic.com/article/135684/declare-war-climate-change-mobilize-wwii (accessed August 18, 2019).
118 Klein, *A Call to Arms*, 17.

119 Goodwin, *No Ordinary Time*, 295.
120 Harvard Thinks Big, "Daniel Gilbert: Global Warming and Psychology," *Vimeo* video, https://vimeo.com/10324258 (accessed August 12, 2019).
121 Jean Carlomusto, dir., *Larry Kramer: In Love & Anger* (HBO, 2015).
122 John Leland, "Twilight of a Difficult Man: Larry Kramer and the Birth of AIDS Activism," *The New York Times*, May 19, 2017, https://www.nytimes.com/2017/05/19/nyregion/larry-kramer-and-the-birth-of-aids-activism.html (accessed August 3, 2019).
123 Larry Kramer, "1,112 and Counting," *New York Native* 59 (March 14–27, 1983).
124 Deborah B. Gould, *Moving Politics: Emotion and Act Up's Fight Against AIDS* (Chicago: University of Chicago Press, 2009).
125 Mark Engler and Paul Engler, *This is an Uprising: How Nonviolent Revolt is Shaping the Twenty-First Century* (New York: Nation Books, 2016).
126 Ibid.
127 David France, T. Woody Richman, Derek Weisehahn, et al., *How to Survive a Plague* (New York: Sundance Selects, 2013).
128 *A National Aids Treatment Research Agenda* (New York: ACT UP, 1989), https://www.poz.com/pdfs/national-aids-treatment-research-agenda-1989.pdf.
129 Nurith Aizenman, "How To Demand A Medical Breakthrough: Lessons From The AIDS Fight," National Public Radio, February 9, 2019, https://www.npr.org/sections/health-shots/2019/02/09/689924838/how-to-demand-a-medical-breakthrough-lessons-from-the-aids-fight (accessed August 18,2019).
130 Albert Jonsen and Jeff Stryker, eds., *The Social Impact of AIDS in the United States* (Washington, D.C.: National Academy Press, 1993), http://www.nap.edu/openbook.php?record_id=1881.
131 UNAIDS, "Fact Sheet — Global AIDS Update 2019," https://www.unaids.org/sites/default/files/media_asset/UNAIDS_FactSheet_en.pdf (accessed August 12, 2019).
132 U.S. Department of Health and Human Services, "Global Statistics," HIV.org, https://www.hiv.gov/hiv-basics/overview/data-and-trends/global-statistics (accessed October 23, 2019).
133 J.R. Vernon, "World War II Fiscal Policies and the End of the Great Depression," *The Journal of Economic History* 54, 4 (Dec. 1994): 850–868.
134 Tom Rosentiel, "How a Different America Responded to the Great Depression," *Pew Research Center*, December 14, 2010, https://

www.pewresearch.org/2010/12/14/how-a-different-america-responded-to-the-great-depression/ (accessed August 18, 2019).

135 "Presidential Approval Ratings — Donald Trump," Gallup, https://news.gallup.com/poll/203198/presidential-approval-ratings-donald-trump.aspx (accessed August 18, 2019).

136 Pew Research Center, "Public Trust in Government: 1958–2019," April 11, 2019, https://www.people-press.org/2019/04/11/public-trust-in-government-1958-2019/ (accessed August 18, 2019).

137 Mark Z. Jacobson and Mark A. Delucchi, "Providing All Global Energy with Wind, Water, and Solar Power, Part I: Technologies, Energy Resources, Quantities and Areas of Infrastructure, and Materials," *Energy Policy* 39, 3 (March 2011): 1154–1169, https://doi.org/10.1016/j.enpol.2010.11.040.

138 "U.S. High Speed Rail Network Map," United States High Speed Rail Association, http://www.ushsr.com/ushsrmap.html (accessed August 18, 2019).

139 C.J. Rhodes, "The Imperative for Regenerative Agriculture," *Science Progress* 100, 1 (March 2017): 80–129, 10.3184/003685017X14876775256165.

Step Five: Join the Climate Emergency Movement

140 Florence White Bull, "I've Been Woken," *Awake: The Film,* https://awakethefilm.org/ (accessed August 4, 2019).

141 "100% Clean Energy Bill Launched by Senators and Movement Leaders," 350.org, https://350.org/press-release/100-clean-energy-bill/ (accessed August 18, 2019).

142 Engler and Engler, *This Is an Uprising.*

143 "Energy and Climate," City of Darebin, http://www.darebin.vic.gov.au/en/Darebin-Living/Caring-for-the-environment/EnergyClimate (accessed August 18, 2019).

144 The Climate Mobilization, "Campaign Background," https://www.theclimatemobilization.org/climate-emergency-overview (accessed August 4, 2019).

145 Anthony Leiserowitz, Edward Maibach, Seth Rosenthal, et al., *Climate Change in the American Mind: April 2019* (New Haven, CT: Yale Program on Climate Change Communication, 2019): doi.org/10.17605/OSF.IO/CJ2NS.

146 Ibid, 17.

147 Abel Gustafson, Anthony Leiserowitz, and Edward Maibach, "Americans Are Increasingly 'Alarmed' about Global Warming," *Yale*

Program on Climate Change Communication, February 12, 2019, https://climatecommunication.yale.edu/publications/americans-are-increasingly-alarmed-about-global-warming/ (accessed October 24, 2019).

148 Mark U. Edwards, *Printing, Propaganda, and Martin Luther* (Berkeley, CA: University of California Press, 1994).

149 Alexis Madrigal, "When the Revolution Was Televised," *The Atlantic*, April 1, 2018, https://www.theatlantic.com/technology/archive/2018/04/televisions-civil-rights-revolution/554639/ (accessed August 18, 2019).

150 Sasha Torres, *Black, White, and in Color: Television and Black Civil Rights* (Princeton, New Jersey: Princeton University Press, 2003).

151 Jane McAlevey, *Raising Expectations (and Raising Hell): My Decade Fighting for the Labor Movement* (New York: Verso, 2014).

152 Becky Bond and Zack Exley, *Rules for Revolutionaries: How Big Organizing Can Change Everything* (White River Junction: Chelsea Green Publishing, 2016).

153 Kimberley Bobo, Jackie Kendall, and Steve Max, *Organizing for Social Change: Midwest Academy for Activists*, 4th ed. (Newport Beach, CA: Seven Locks Press, 2010).

154 Paul Engler, "Protest Movements Need the Funding They Deserve," *Stanford Social Innovation Review*, July 3, 2018 (accessed August 12, 2019).

155 Kim Klein, *Fundraising for Social Change* (San Francisco: Jossey-Bass, 2011).

Conclusion: Live as a Climate Warrior

156 Abraham Maslow, "A theory of human motivation," *Psychological Review*, 50 (4), 370–396, http://dx.doi.org/10.1037/h0054346.

157 Abraham Maslow and Rober Frager, *Motivation and Personality* (New Delhi: Pearson Education, 2007).

158 Kim Polman and Stephen Vasconcellos-Sharpe, *Imaginal Cells; Visions of Transformation* (London: Reboot the Future, 2017).

WORKS CITED

350.org. "100% Clean Energy Bill Launched by Senators and Movement Leaders." 350.org, April 27, 2017. https://350.org/press-release/100-clean-energy-bill/.

"A Field Guide to the U.S. Environmental Movement." *Inside Climate News,* April 8, 2015. https://insideclimatenews.org/content/field-guide-us-environmental-movement.

ACT UP. *A National AIDS Treatment Research Agenda.* New York: ACT UP, 1989. https://www.poz.com/pdfs/national-aids-treatment-research-agenda-1989.pdf.

Aizenman, Nurith. "How To Demand A Medical Breakthrough: Lessons From The AIDS Fight." *National Public Radio,* February 9, 2019. https://www.npr.org/sections/health-shots/2019/02/09/689924838/how-to-demand-a-medical-breakthrough-lessons-from-the-aids-fight.

Baker, Carolyn, and Guy R. McPherson. *Extinction Dialogues: How to Live with Death in Mind.* Oakland: Next Revelation Press, 2014.

Baker, Carolyn. *Collapsing Consciously: Transformative Truths for Turbulent Times.* Berkeley: North Atlantic Books, 2013.

Banerjee, Neela, Lisa Song, and David Hasemyer. "Exxon's Own Research Confirmed Fossil Fuels' Role in Global Warming Decades Ago." *Inside Climate News,* September 16, 2015. https://insideclimatenews.org/news/15092015/Exxons-own-research-confirmed-fossil-fuels-role-in-global-warming.

Bassett, Thomas. "Reaping on the Margins: A Century of Community Gardening in America." *Landscape* 25, 2 (1981): 1–8.

Bobo, Kimberley. *Organizing for Social Change: Midwest Academy for Activists,* 4th ed. Newport Beach: Seven Locks Press, 2010.

Bond, Becky, and Zack Exley. *Rules for Revolutionaries: How Big Organizing Can Change Everything.* White River Junction: Chelsea Green Publishing, 2016.

Brach, Tara. "RAIN: Recognize, Allow, Investigate, Nurture." *Tara Brach* (blog). https://www.tarabrach.com/rain/.

Brown, Lester. "Could Food Shortages Bring Down Civilization?" *Scientific American* (May 2009): 50–57. http://www.earth-policy.org/images/uploads/press_room/SciAm-final.pdf.

----. Plan B 4.0: *Mobilizing to Save Civilization.* New York: W.W. Norton, 2009.

Brulle, Robert J. "The Climate Lobby: A Sectoral Analysis of Lobbying Spending on Climate Change in the USA, 2000 to 2016." *Climatic Change: An Interdisciplinary, International Journal Devoted to the Description, Causes and Implications of Climatic Change* 149, 3–4 (2018): 289–303.

----. "Institutionalizing Delay: Foundation Funding and the Creation of U.S. Climate Change Counter-Movement Organizations." *Climatic Change: An Interdisciplinary, International Journal Devoted to the Description, Causes and Implications of Climatic Change* 122, 4 (2014): 681–694.

Bull, Florence White. "I've Been Woken." *Awake: A Dream from Standing Rock* (website), https://awakethefilm.org/.

Carlomusto, Jean. Larry Kramer: *In Love & Anger.* HBO Documentaries, 2015. https://www.hbo.com/documentaries/larry-kramer-in-love-and-anger.

Center for Biological Diversity. "The Extinction Crisis." https://www.biologicaldiversity.org/programs/biodiversity/elements_of_biodiversity/extinction_crisis/.

Chait, Jonathan. "Why Are the Republicans the Only Climate-Science Denying Party in the World?" *New York Magazine,* September 27, 2015. http://nymag.com/intelligencer/2015/09/whys-gop-only-science-denying-party-on-earth.html?gtm=top.

Chapman, Daniel, Brian Lickel, and Ezra Markowitz. "Reassessing Emotion in Climate Change Communication." *Nature* 7 (2017): 850–852.

Cialdini, Robert B. *Influence: Science and Practice,* 5th ed. Harlow, Essex: Pearson, 2008.

City of Darebin. "Energy and Climate." http://www.darebin.vic.gov.au/en/Darebin-Living/Caring-for-the-environment/EnergyClimate.

The Climate Mobilization. "An In-Depth Look at the Climate Emergency Movement: Campaign Background." https://www.theclimatemobilization.org/climate-emergency-overview.

DARA and the Climate Vulnerable Forum. *Climate Vulnerability Monitor: A Guide to the Cold Calculus of a Hot Planet,* 2nd ed. Spain: DARA, 2012. https://daraint.org/wp-content/uploads/2012/09/CVM2ndEd-FrontMatter.pdf.

Darby, Megan. "Meet the Woman Who First Identified the Greenhouse Effect." *Climate Change News,* February 9, 2016. https://www.climatechangenews.com/2016/09/02/the-woman-who-identified-the-greenhouse-effect-years-before-tyndall/.

Darley, John M., and Bibb Latané. "Bystander Intervention in Emergencies: Diffusion of Responsibility." *Journal of Personality and Social Psychology* 8, 4 (1968): 377–83.

Duberstein, John. *A Velvet Revolution: Václav Havel and the Fall of Communism*. Greensboro: Morgan Reynolds, 2006.

Eboch, M.M. *Native American Code Talkers*. Minneapolis: ABDO, 2015.

Edwards, Mark U. *Printing, Propaganda, and Martin Luther*. Berkeley: University of California Press, 1994.

Engler, Mark, and Paul Engler. *This is an Uprising: How Nonviolent Revolt Is Shaping the Twenty-First Century*. New York: Nation Books, 2016.

Engler, Paul. "Protest Movements Need the Funding They Deserve." *Stanford Social Innovation Review*, July 3, 2018.

European Commission. "Worrying Effects of Accelerating Climate Change on the Mediterranean Basin." *EU Science Hub*, October 22, 2019. https://ec.europa.eu/jrc/en/science-update/worrying-effects-accelerating-climate-change-mediterranean-basin.

Fishback, Price V., and Joseph Cullen. *Did Big Government's Largesse Help the Locals? The Implications of WWII Spending for Local Economic Activity, 1939–1958*. Cambridge: National Bureau of Economic Research, 2006.

Frey, William H., and Muriel Langseth. *Crying: The Mystery of Tears*. Minneapolis: Winston Press, 1985.

Friedman, Thomas. *Hot, Flat, and Crowded: Why We Need a Green Revolution — and How It Can Renew America*. New York: Farrar, Straus, and Giroux, 2008.

Fromm, Erich. "Credo." *Erich Fromm Online*, https://fromm-online.org/en/das-leben-erich-fromms/fromms-credo-eines-humanisten/.

----. "Love of Death and Love of Life." *The Heart of Man: Its Genius for Good and Evil*. Edited by Ruth Nanda Anshen. New York: Harper, 1964.

Gallup. "Presidential Approval Ratings — Donald Trump." *Gallup Presidential Approval Ratings*. https://news.gallup.com/poll/203198/presidential-approval-ratings-donald-trump.aspx.

Gaind, Nisha. "Wildlife in Decline: Earth's Vertebrates Fall 58% in Past Four Decades." *Nature*, October 28, 2016. https://www.nature.com/news/wildlife-in-decline-earth-s-vertebrates-fall-58-in-past-four-decades-1.20898.

Giridharandas, Anand. *Winners Take All: The Elite Charade of Changing the World*. New York: Knopf, 2018.

Gleick, Peter H. "Water, Drought, Climate Change, and Conflict in Syria." *Weather, Climate, and Society* 6, 3 (2014): 331–34.

Goodwin, Doris. "The Way We Won: America's Economic Breakthrough During WWII." *American Prospect,* Fall 1992. https://prospect.org/article/way-we-won-americas-economic-breakthrough-during-world-war-ii.

Goodwin, Doris Kearns. *No Ordinary Time: Franklin & Eleanor Roosevelt: The Home Front in WW II.* New York: Simon & Schuster, 1994.

Gould, Deborah B. *Moving Politics: Emotion and ACT UP's Fight Against AIDS.* Chicago: University of Chicago Press, 2009.

Gračanin, Asmir, Lauren Bylsma, and Ad J.J.M Vingerhoets. "Is Crying a Self-Soothing Behavior?" *Frontiers in Psychology* 5 (May 2014), doi: 10.3389/fpsyg.2014.00502.

Grooten, M., and R.E.A. Almond, eds. *Living Planet Report 2018: Aiming Higher.* Switzerland: WWF, 2018. https://wwf.panda.org/knowledge_hub/all_publications/living_planet_report_2018/.

Guenther, Genevieve. "Who Is the 'We' in 'We Are Causing Climate Change'?" *Grist,* October 13, 2018. https://grist.org/article/who-is-the-we-in-we-are-causing-climate-change/.

Hallman, Caspar, Martin Sorg, Eelke Jongejans, Henk Siepel, Nick Hofland, Heinz Schwan, Werner Stenmans, et al. "More than 75 Percent Decline Over 27 Years in Total Flying Insect Biomass in Protected Areas." *PloS ONE* 12, 10 (2017): https://doi.org/10.1371/journal.pone.0185809.

Harvard Thinks Big. "Daniel Gilbert: Global Warming and Psychology." *Vimeo,* video. 11:19, https://vimeo.com/10324258.

Havel, Václav. *The Power of the Powerless: Citizens Against the State in Central Eastern Europe.* Translated by John Keane. Abingdon: Routledge, 1985.

Heglar, Mary Annaïse. "The Big Lie We're Told about Climate Change Is that It's Our Fault." *Vox,* November 27, 2018. https://www.vox.com/first-person/2018/10/11/17963772/climate-change-global-warming-natural-disasters.

Herman, Arthur. *Freedom's Forge: How American Business Produced Victory in World War II.* New York: Random House, 2013.

How to Survive a Plague. Directed by David France. New York: Public Square Films, 2012.

IPCC. *Global Warming of 1.5°C: An IPCC Special Report on the Impacts of Global Warming of 1.5°C Above Pre-Industrial Levels and Related Global Greenhouse Gas Emission Pathways.* Edited by V. Masson-Delmotte, P. Zhai, H.O. Pörtner, D. Roberts, J. Skea, P.R. Shukla, A. Pirani et al. In Press, 2018. https://www.ipcc.ch/sr15/.

Jacobson, Mark Z., and Mark A. Delucchi. "Providing All Global Energy with Wind, Water, and Solar Power, Part I: Technologies, Energy Resources, Quantities and Areas of Infrastructure, and Materials." *Energy Policy* 39, 3 (March 2011): 1154–1169. https://doi.org/10.1016/j.enpol.2010.11.040.

Jonsen, Albert, and Jeff Stryker, eds. *The Social Impact of AIDS in the United States*. Washington, D.C.: National Academy Press, 1993. http://www.nap.edu/openbook.php?record_id=1881.

Kaiser, David. *No End Save Victory: How FDR Led the Nation into War.* New York: Basic Books, 2015.

Keay, Douglas. "Interview for *Women's Own*." Margaret Thatcher Foundation, October 16, 1984. https://www.margaretthatcher.org/document/105577.

Keller, Jared. "The U.S. Suicide Rate Is at Its Highest in Half a Century." *Pacific Standard*, December 4, 2018. https://psmag.com/news/the-suicide-rate-is-at-its-highest-in-a-half-century.

Kersten, Andrew Edmund. *Race, Jobs, and the War: The FEPC in the Midwest, 1941–46*. Urbana, IL: University of Illinois Press, 2007.

Kimble, James J. *Mobilizing the Home Front: War Bonds and Domestic Propaganda*. College Station, TX: Texas A&M University Press, 2006.

King, Martin Luther, Jr. *Letter from a Birmingham Jail*, August 1963. http://web.cn.edu/kwheeler/documents/letter_birmingham_jail.pdf.

Klein, Kim. *Fundraising for Social Change*. San Francisco: Jossey-Bass, 2011.

Klein, Maury. *A Call to Arms: Mobilizing America for World War II.* New York: Bloomsbury Publishing, 2013.

Klein, Naomi. *This Changes Everything: Capitalism vs. The Climate*. New York: Simon & Schuster, 2014.

Kramer, Larry. "1,112 and Counting." *New York Native* 59 (March 14–27, 1983).

Latané, Bibb, and John M. Darley. *Group Inhibition of Bystander Intervention in Emergencies*. Emmitsburg, MD: National Emergency Training Center, 1968.

"Leaders: A Greener Bush." *The Economist*, February 15, 2003. https://www.economist.com/leaders/2003/02/13/a-greener-bush.

Leeuwen, N. van, E.R. Bossema, H. van Middendorp, A.A. Kruize, H. Bootsma, J.W.J. Bijlsma, and R. Geenen. "Dealing with Emotions When the Ability to Cry is Hampered: Emotion Processing and Regulation in Patients with Primary Sjogren's Syndrome." *Clinical and Experimental Rheumatology* 30, 4 (2012): 492–498. http://hdl.handle.net/2066/110788.

Leiserowitz, Anthony, Edward Maibach, Seth Rosenthal, John Kotcher, Parrish Bergquist, Matthew Ballew, Matthew Goldberg et al. *Climate Change in the American Mind: April 2019*. New Haven, CT: Yale Program on Climate Change Communication, 2019. doi.org/10.17605/OSF.IO/CJ2NS.

Leland, John. "Twilight of a Difficult Man: Larry Kramer and the Birth of AIDS Activism." *The New York Times*, May 19, 2017. https://www.nytimes.com/2017/05/19/nyregion/larry-kramer-and-the-birth-of-aids-activism.html.

Macy, Joanna, and Chris Johnstone. *Active Hope: How to Face the Mess We're In Without Going Crazy*. San Francisco: New World Library, 2012.

Madrigal, Alexis. "When the Revolution Was Televised." *The Atlantic*, April 1, 2018. https://www.theatlantic.com/technology/archive/2018/04/televisions-civil-rights-revolution/554639/.

Mann, Michael, Susan Hassol, and Tom Tole. "Doomsday Scenarios Are as Harmful as Climate Change Denial." *The Washington Post*, July 12, 2017. https://www.washingtonpost.com/opinions/doomsday-scenarios-are-as-harmful-as-climate-change-denial/2017/07/12/880ed002-6714-11e7-a1d7-9a32c91c6f40_story.html?utm_term=.c6bff330726d.

Maslow, Abraham. "A Theory of Human Motivation." *Psychological Review* 50, 4 (1943) 370–396. http://dx.doi.org/10.1037/h0054346.

Maslow, Abraham, and Robert Frager. *Motivation and Personality*. New Delhi: Pearson Education, 2007.

McAlevey, Jane. *Raising Expectations (and Raising Hell): My Decade Fighting for the Labor Movement*. New York: Verso, 2014.

McKibben, Bill. "A World At War." *The New Republic*, August 15, 2016. https://newrepublic.com/article/135684/declare-war-climate-change-mobilize-wwii.

Moore, Thomas, and Donald Mattison. "Adult Utilization of Psychiatric Drugs and Differences by Sex, Age, and Race." *JAMA Internal Medicine* 177 (February 2017): 274–275. doi:10.1001/jamainternmed.2016.7507.

National Safety Council. "Preventable Deaths." *Injury Facts*, 2018. https://injuryfacts.nsc.org/all-injuries/preventable-death-overview/odds-of-dying/data-details/.

National Science Foundation. *The National Science Foundation: A Brief History*. Washington, D.C.: ERIC Clearinghouse, 1988.

Neff, Kristen. *Self-Compassion: The Proven Power of Being Kind to Yourself*. New York: HarperCollins, 2011.

Nielsen. "The Total Audience Report: Q1 2016." June 27, 2016. https://www.nielsen.com/us/en/insights/report/2016/the-total-audience-report-q1-2016/.

Nilsson, D., and A. Johansson. "Social Influence During the Initial Phase of a Fire Evacuation — Analysis of Evacuation Experiments in a Cinema Theatre." *Fire Safety Journal* 44 (2008): 71–79.

Norwegian Ministry of Trade, Industry and Fisheries. *The State Ownership Report 2015*. https://www.equinor.com/content/dam/statoil/documents/the-state-ownership-report-2015.pdf.

Novick, Kerry Kelly, and Jack Novick. *Emotional Muscle: Strong Parents, Strong Children*. Bloomington, IN: Xlibris, 2010.

Oreskes, Naomi, and Erik Conway. *Merchants of Doubt: How a Handful of Scientists Obscured the Truth on Issues from Tobacco Smoke to Global Warming*. New York: Bloomsbury Publishing, 2010.

Panksepp, Jaak. "Affective Neuroscience of the Emotional BrainMind: Evolutionary Perspectives and Implications for Understanding Depression." *Dialogues in Clinical Neuroscience* 12, 4 (December 2010): 533–45.

Peterson, D.J. *Troubled Lands: The Legacy of Soviet Environmental Destruction*. Boulder: Westview Press, 1993.

Pew Research Center. "Public Trust in Government: 1958–2019." *Pew Research Center, U.S. Politics & Policy*, April 11, 2019. https://www.people-press.org/2019/04/11/public-trust-in-government-1958-2019/.

Picchi, Aimee. "A Surprise Expense Would Put Most Americans into Debt." *CBSNews.com*, January 12, 2017. https://www.cbsnews.com/news/most-americans-cant-afford-a-500-emergency-expense/.

Polman, Kim, and Stephen Vasconcellos-Sharpe. *Imaginal Cells; Visions of Transformation*. London: Reboot the Future, 2017.

Rhodes, C.J. "The Imperative for Regenerative Agriculture." *Science Progress* 100, 1 (March 2017): 80–129, 10.3184/003685017X14876775256165.

Richardson, Valerie. "Obama Takes Credit for U.S. Oil-and-Gas Boom: 'That Was Me, People.'" *The Washington Times*, November 28, 2018. https://www.washingtontimes.com/news/2018/nov/28/obama-takes-credit-us-oil-and-gas-boom-was-me-peop/.

Rockoff, Hugh. *Keep on Scrapping the Salvage Drives of World War II*. Cambridge: National Bureau of Economic Research, 2007. http://papers.nber.org/papers/w13418.

Romm, Joe. "Ocasio-Cortez Says We Need World War II-Scale Action on Climate: Here's What that Looks Like." *Think Progress*, February 12, 2019.

https://thinkprogress.org/green-new-deal-ocasio-cortez-7c9ac944b37d/.

Rosentiel, Tom. "How a Different America Responded to the Great Depression." *Pew Research Center,* December 14, 2010. https://www.pewresearch.org/2010/12/14/how-a-different-america-responded-to-the-great-depression/.

Rottenberg, J., F.H. Wilehelm, J.J. Gross, and I.H. Gotlib. "Vagal Rebound During Resolution of Tearful Crying Among Depressed and Nondepressed Individuals." *Psychophysiology* 40 (2003): 1–6

Saad, Lydia. "Gallup Vault: A Country Unified After Pearl Harbor." *Gallup,* December 5, 2016. https://news.gallup.com/vault/199049/gallup-vault-country-unified-pearl-harbor.aspx.

Scherer, Glenn. "Climate Science Predictions Prove Too Conservative." *Scientific American,* December 6, 2012, https://www.scientificamerican.com/article/climate-science-predictions-prove-too-conservative/.

Schultz, Cheryl, Leone Brown, Emma Pelton, and Elizabeth Crone. "Citizen Science Monitoring Demonstrates Dramatic Declines of Monarch Butterflies in Western North America." *Biological Conservation* 214 (2017): 343–346. https://doi.org/10.1016/j.biocon.2017.08.019.

Seidenkrantz, Marit-Solveig. "80 Years Since the First Calculations Showed that the Earth Was Warming Due to Rising Greenhouse Gas Emissions." *Science X,* June 5, 2018. https://phys.org/news/2018-06-years-earth-due-greenhouse-gas.html (accessed July 24, 2019).

Sherwood, Harriet, and Angela Guiffrida. "Pope Reveals He Had Weekly Psychoanalysis Sessions at Age 42." *The Guardian,* September 1, 2017. https://www.theguardian.com/world/2017/sep/01/pope-francis-psychoanalysis.

Shome, Debika, and Sabine Marx. *The Psychology of Climate Change Communication: A Guide for Scientists, Journalists, Educators, Political Aides, and the Interested Public.* New York: Center for Research on Environmental Decisions, 2009.

Spratt, David, and Ian Dunlop. *Existential Climate Risk: A Scenario Approach.* Melbourne, Australia: Breakthrough, 2019. https://docs.wixstatic.com/ugd/148cb0_a1406e0143ac4c469196d3003bc1e687.pdf.

———. *What Lies Beneath: The Understatement of Existential Climate Risk.* Melbourne, Australia: Breakthrough, 2018. https://docs.wixstatic.com/ugd/148cb0_a1406e0143ac4c469196d3003bc1e687.pdf.

Spratt, David, and Philip Sutton. *Climate Code Red: The Case for Emergency Action.* https://www.amazon.com/Climate-Code-Red-Emergency-

Action-ebook/dp/B004C05HIY Scribe Publications Pty Ltd. (November 13, 2010) April 1, 2009

Steimer, Thierry. "The Biology of Fear- and Anxiety-Related Behaviors." *Dialogues in Clinical Neuroscience* 4, 3 (September 2002): 231–49.

Stiglitz, Joe. "The Climate Crisis Is Our Third World War: It Needs a Bold Response." *The Guardian,* June 4, 2019. https://www.theguardian.com/commentisfree/2019/jun/04/climate-change-world-war-iii-green-new-deal.

Stockholm Resilience Centre. "Planetary Boundaries Research." https://www.stockholmresilience.org/research/planetary-boundaries.html.

Takaki, Ronald T. *Double Victory: A Multicultural History of America in World War II.* Boston: Little, Brown and Company, 2001.

Tobis, Michael, and Stephen Ban. "OK, Getting Serious Again." *Only in It for the Gold* (blog), January 14, 2010. http://init.planet3.org/2010/01/ok-getting-serious-again.html.

Torres, Sasha. *Black, White, and in Color: Television and Black Civil Rights.* Princeton, New Jersey: Princeton University Press, 2003.

UNAIDS. "Fact Sheet — Global AIDS Update 2019." UNAIDS, https://www.unaids.org/sites/default/files/media_asset/UNAIDS_FactSheet_en.pdf.

United Nations Security Council. "Statement by the President of the Security Council." United Nations, January 30, 2018. https://www.un.org/en/ga/search/view_doc.asp?symbol=S/PRST/2018/3.

United Planet Faith & Science Initiative, "Greta Thunberg, 'Our House Is on Fire.' 2019 World Economic Forum (WEF) in Davos." YouTube video, 6:03. January 25, 2019. https://www.youtube.com/watch?v=zrF1THd4bUM.

Vernon, J. R. "World War II Fiscal Policies and the End of the Great Depression." *The Journal of Economic History* 54, 4 (December 1994): 850–868.

Vingerhoets, A.J.J.M., N. van de Ven and Y. van der Velden. "The Social Impact of Emotional Tears." *Motivation and Emotion* 40, 3 (2016): 455–463.

Vingerhoets, A.J.J.M. and M.J.J. Wubben. "The Health Benefits of Crying." *Emotion Researcher* 23, 1 (2008): 15–17.

Wallace Wells, David. *The Uninhabitable Earth: Life After Warming.* New York: Penguin Random House, 2019.

White House Office of Management and Budget. "Table 3.1." *Historical Tables: Budget of the United States Government Fiscal Year 2011.* https://www.whitehouse.gov/omb/historical-tables/.

Wilson, Edward O. *Half-Earth: Our Planet's Fight for Life.* New York: Liveright, 2016.
----. *The Social Conquest of Earth.* New York: Liveright, 2012.

ABOUT THE AUTHORS

MARGARET KLEIN SALAMON, PHD, is a clinical psychologist turned climate warrior whose work helps people to face the deeply frightening, painful truths of climate change and transform their despair into effective action.

Though she loved being a therapist, in the aftermath of Hurricane Sandy she woke up to climate truth and founded The Climate Mobilization and Climate Mobilization Project. These organizations have led a campaign that has sparked thousands of Climate Emergency Declarations by governments around the world and played a key role in bringing the framing of WWII-scale mobilization into U.S. politics. She is the author of *Leading the Public into Emergency Mode* and *The Transformative Power of Climate Truth.*

Born and raised in Ann Arbor, Michigan, Margaret earned her PhD in clinical psychology from Adelphi University and holds a BA in social anthropology from Harvard. She now lives in Brooklyn, NY, with her husband, Bren. To read more, share your story, and receive ongoing updates, visit:

www.TheClimateMobilization.org/Updates

MOLLY GAGE, PHD, is committed to developing nonfiction books that push forward progressive ideas and elevate the voices that think them.

ABOUT THE ORGANIZATION

THE CLIMATE MOBILIZATION and Climate Mobilization Project are working to initiate a WWII-scale emergency mobilization to protect humanity and the living world from climate catastrophe.

TCM is an innovation lab that develops and pioneers the strongest and most aggressive policy frameworks, messaging, and organizing approaches to reverse global warming and accelerate the global transition into "emergency mode." Prior to TCM's founding in 2014, there was no climate group organizing around the need for WWII-scale emergency speed transition to a zero-emissions just economy. At the time of this printing, thousands of governments, representing over 800 million people, have adopted TCM's core framing by declaring a Climate Emergency.

To learn more, volunteer, make a donation, or start a local Climate Emergency campaign, please visit www.TheClimateMobilization.org.

The author's proceeds from the sale of this book are being donated to Climate Mobilization Project, a 501(c)(3) nonprofit organization affiliated with The Climate Mobilization. You are also invited to support this work with a tax-deductible gift by visiting www.TheClimateMobilization.org/Donation

A Note about the Publisher

New Society Publishers is an activist, solutions-oriented publisher focused on publishing books for a world of change. Our books offer tips, tools, and insights from leading experts in sustainable building, homesteading, climate change, environment, conscientious commerce, renewable energy, and more — positive solutions for troubled times.

We're proud to hold to the highest environmental and social standards of any publisher in North America. When you buy New Society books, you are part of the solution!

- We print all our books in North America, never overseas
- All our books are printed on **100% post-consumer recycled paper**, processed chlorine free, with low-VOC vegetable-based inks (since 2002)
- Our corporate structure is an innovative employee shareholder agreement, so we're one-third employee-owned (since 2015)
- We're carbon-neutral (since 2006)
- We're certified as a B Corporation (since 2016)

At New Society Publishers, we care deeply about *what* we publish — but also about *how* we do business.

Download our catalogue at https://newsociety.com/Our-Catalog, or for a printed copy please email info@newsocietypub.com or call 1-800-567-6772 ext 111.

New Society Publishers
ENVIRONMENTAL BENEFITS STATEMENT

*By Using 100% post-consumer recycled paper vs virgin paper stock, New Society Publishers saves the following resources:[1] (per every 5,000 copies printed)

30	Trees
2,695	Pounds of Solid Waste
2,965	Gallons of Water
3,867	Kilowatt Hours of Electricity
4,899	Pounds of Greenhouse Gases
12	Pounds of HAPs, VOCs, and AOX Combined
7	Cubic Yards of Landfill Space

[1]Environmental benefits are calculated based on research done by the Environmental Defense Fund and other members of the Paper Task Force who study the environmental impacts of the paper industry.